Management Division 2017

Core Programming Area at the 2017 AIChE Annual Meeting

Minneapolis, Minnesota, USA
29 October – 3 November 2017

ISBN: 978-1-5108-5794-0

Printed from e-media with permission by:

Curran Associates, Inc.
57 Morehouse Lane
Red Hook, NY 12571

Some format issues inherent in the e-media version may also appear in this print version.

Printed by Curran Associates, Inc. (2018)

For permission requests, please contact AIChE
at the address below.

AIChE
120 Wall Street, FL 23
New York, NY 10005-4020

Phone: (800) 242-4363
Fax: (203) 775-5177

www.aiche.org

Additional copies of this publication are available from:

Curran Associates, Inc.
57 Morehouse Lane
Red Hook, NY 12571 USA
Phone: 845-758-0400
Fax: 845-758-2633
Email: curran@proceedings.com
Web: www.proceedings.com

TABLE OF CONTENTS

Some abstracts in this book refer to images that were not made available by the publisher

Navigating the Unpaved Roads and Knowing the Unwritten Rules: Advancement for Teaching-Focused Faculty (Abstract ID# 483241)

Lisa G. Bullard
North Carolina State University

Unlike the tenure track faculty ranks where mentors are routinely assigned and the requirements to obtain tenure are clearly articulated, teaching-focused faculty face a landscape in which there may not be a senior faculty member to serve as a mentor, and the path to promotion may not be clearly defined or defined at all. Departmental voting faculty may not understand how to evaluate the promotion dossier for a teaching-focused faculty member. Departmental resources for professional development may be limited, and the funding landscape is less certain. The good news is that there are travelers who have walked this way before and established some signposts along the way. This paper will provide practical suggestions for teaching-focused faculty who would like to hone their teaching and educational scholarship skills, demonstrate leadership and contribution on- and off-campus, and obtain recognition for their work.

What each institution calls these individuals may vary greatly; some of these might include "teaching professors", "lecturers", "professors of the practice", "non-tenure stream", 'non-tenure track", etc. (For the purposes of this paper, I will refer to the position as a "teaching professor"). Likewise, each university has policies and procedures about promotion and tenure. Make sure that you are aware of those policies and follow them – to the letter. Typically these policies are located on the web site of the Provost or other academic officer. They may also be found on the web site for the Office of Faculty Development or similar organization. These are the "official" documented policies. However, often there are "unwritten" (and sometimes unspoken) policies for promotion in each department. The suggestions below are intended to help non-tenure track faculty best position themselves for contribution, advancement, and promotion. Ready? Let's begin!

Take advantage of opportunities to hone your teaching skills, demonstrate leadership and contribution on- and off-campus, and obtain recognition for your work.

- Find ways to get in front of your faculty and make them aware of your "high impact" contributions – make presentations at faculty meetings or workshops, industrial advisory board meetings, conduct teaching/advising tips workshops within the department.
- Attend on- and off-campus teaching workshops. At some point, you may want to start teaching such workshops yourself!
- Professional societies offer regional and national workshops as well as opportunities for leadership in specific divisions of interest. There may be other relevant professional societies in which you can become involved. Consider running for office or volunteering to chair a committee either at the regional or national level.
- Make conference presentations related to your teaching and/or educational research at regional and/or national conferences. Look closely at what you are already doing that you can share – it doesn't have to be "extra" work.
- Identify relevant educational or trade journals related to your field or area of teaching and consider submitting an article. Some journals accept one-page "Teaching Tips" which can be a good way to get started with a good idea to share that doesn't require a lot of assessment or funding.
- Participate in and take a leadership role in course and curriculum committees, coordinators of advising committees, accreditation teams, etc.
- Apply or ask to be nominated for teaching and advising awards as appropriate. If there is an award that is "expected" for faculty in your role, be intentional about examining the criteria for the award and determining what you need to do to be a strong candidate.
- Find out about TA training for your department or college and offer to present a workshop or participate in an existing training session.
- Consider participation in multidisciplinary or multi-institution research studies in your area of interest or teaching specialty. (Conferences are a great way to meet others who teach in your area or share similar interests).
- Move forward best educational practices across your department and college. Take initiative in promoting teaching effectiveness efforts on campus and within your department.
- Work with your Department Head to become an associate (or full) member of the graduate faculty. This is needed for NSF and other funding applications.

- Identify potential co-advising opportunities for PhD students -- those in discipline areas or those for students are interested in educational sides of fields. This is especially appropriate if being co-PI on a grant will lead to publications and external visibility.
- Negotiate with your department to pay for one conference a year, with the possibility for two if you have a paper published or presented. You might also ask for a laptop in your first year until you get funds to replace it several years in. AIChE and ASEE provide excellent opportunities to present your work and network with others.
- Promote your work (teaching, advising, research, extension, engagement) on Facebook, Twitter, the departmental web site, etc. College publications are always looking for good stories. Don't be too humble!
- Make sure that you are an active and visible part of departmental life. While you may not be part of the DVF, you should still have "a seat at the table" when discussions are held about courses, curricula, faculty hiring, and departmental governance. Make sure that you are included on email communications within the department. Actively participate in faculty meetings and make sure that you are listed in departmental materials (web site, recruiting materials, etc.)

Apply for grants for educational or disciplinary research

- Seek out interactions with potential funding agencies; face-to-face interactions with program directions are invaluable in building relationships and networking.
- Volunteer to be on a review panel by sending a 1-2 page CV to the Director. This is a great way to see both successful and unsuccessful proposals and make contacts/connections in your field.
- Make sure you're submitting your proposal to the right division – call the program director and discuss before spending a lot of time targeting the wrong solicitation.
- Prepare a one-page white paper as a discussion tool before generating a long document.
- Get connected in the "broader impact" area of other proposals. Make sure that the PI lists you as a co-PI or includes funding for your work, funding for your travel to conferences, or some type of recognition/compensation.

Manage Your Statement of Mutual Expectations

- Maintain a *Statement of Mutual Expectations* (SME). Depending on your institution, the name of this document may be different, but it typically documents the mutually-agreed-upon expectations from both you and the department during your appointment. It is a written description of the mix of your realms of responsibility – teaching, research, and extension. It is typically prepared during your first year and is maintained in your personnel folder through your course of service to the University.

- The SME should be reviewed annually with your Department Head, and it should be changed when necessary, e.g. after promotion and when changes to departmental leadership occur.
- You should discuss with your Department Head how your realms of responsibility may need to change if you plan to pursue promotion. For example, it's difficult to be engaged with external collaborators, do educational research, and demonstrate external impact if you are teaching multiple sections of large class each semester. Any changes need to be documented and dated.
- Evaluation for promotion consists of (a) SME (your assignment), (b) departmental rules (the level of quality expected) and (c) your annual review/assessment (are you meeting these expectations?) You should be getting feedback from your Department Head every year. If not – ask.
- Ultimately it is your responsibility to make sure that you are not working on a regular basis outside the realms articulated in your SME. An occasional activity is not a problem, but a regular assignment in a realm outside the SME may raise a red flag. If this is the case, then your SME may need updating.
- Explore your department's or college's strategic plan and put information in the SME that reflects the mission/vision/goals and is aligned with what you want to do now or in the future. The SME can be a starting point for a discussion of what you want your career to look like. It may involve some iteration or negotiation.

Prepare Your Package for Promotion

- Find out what format promotion documents require and begin to document your yearly accomplishments in that format. Number the publications and activities (for those who might just be looking at the numbers).
- Know the policies and regulations that affect your employment at your institution.
- Make sure that you have a formal Peer Evaluation of your teaching on a routine basis, as designated by your department. This is typically a critical component in the review and promotion processes.

- Discuss the issue of promotion with your Department Head at your annual review. Be sure your Department Head knows of your interest to submit your dossier for promotion by the previous spring. Departments typically submit a list of those "going up" in the early fall, when promotion packages are due.
- Request a milestone review by the Departmental Voting Faculty (DVF) in advance of your going up for promotion. This helps the faculty see what you are doing and provides a chance for early feedback.
- Teaching faculty are typically evaluated against the standards of their department and within the context of college and department RPT rules. Talk to your Department Head about his/her understanding of those standards and rules.
- How is "scholarly activity" defined? For teaching-focused faculty, it can be more than what happens in the classroom. This could include service responsibilities, participation/leadership in professional organizations outside the university, attending workshops, sitting on panels, etc. In other words, "talking and thinking about teaching". "Research" may be within your discipline or in the area of teaching and learning. The expectation is that you are "scholarly in whatever you do".
- Do successful candidates have to have something to "show" in every realm of responsibility? No. However, to be eligible for promotion, you do need to show that you are doing the items on the SME really well. Remember that "the SME is not aspirational, it's achievable". Other things can be thrown in (administrative assignments, service, etc.) You may have something in the dossier that is not on the SME – you can still include it.
- If you are the first teaching-focused faculty to go up for promotion, or your department is not familiar with the criteria for teaching faculty promotion, benchmark others in your college or at the university. Ask your Department Head if a teaching faculty member from another Engineering department can review your package and provide input and feedback about your contributions.
- Find someone in the DVF to be your advocate/sponsor. They can speak up for you at promotion time. Identify someone who is influential and a thought shaper in the department, and make sure they know what you're doing and what impact you have had.
- When composing your candidate statement, look at the wording of the promotion criteria *and use those same words.* Make it easy for the committee to see how you have met the criteria.

Create Community

Identify other teaching-focused faculty in your department, college, or university, in addition to faculty at other institutions who have similar roles. Besides being a valuable source of new ideas and approaches, it's encouraging to have similar-minded colleagues with whom you can vent – and celebrate. As a group, you might also be able to influence organizational policy around non-tenured promotion procedures or advocate for resources or training. If there's not such a group in your college or university, start one!

Conclusions

The suggestions in this paper are based on the author's personal experience, as well as the experience of other teaching-focused faculty at North Carolina State University and colleagues at other institutions. The title of this paper, which refers to "unpaved roads and unwritten rules", suggests that while your career path as a non-tenured track faculty member may be less predictable, it is also yours to define. Whether your focus is teaching, educational research, advising, administration, or some combination of these, you play a key role in the department and in the lives of your students. If you follow the advice in this paper, there are no guarantees of promotion, but I predict that you'll feel more empowered in your role and understand what is required to seek promotion, if that is your choice.

503908 Networking for Nerds: How to Land (or Create) Your Dream Job and Keep Your Career Moving Forward!

Abstract:

Wanna land your dream job? Get ready to network! Most jobs and other game-changing career opportunities are not advertised, and even if they are, there is usually a short-list of candidates already in mind. So how do you find out about and access the 90% of jobs and other opportunities that are "hidden"? In this workshop, we will focus on proven networking strategies and tactics to identify new opportunities, locate decision-makers within organizations, solidify your reputation and brand in the minds of those who hire, and gain access to hidden jobs and game-changing opportunities. Discover how networking and self-promotion can enable you to land or even create your dream job from scratch!

509054 Solve this! Fundamental Approach to Problem Solving in Industrial Processes I (Invited Talks)

Abstract:

Solving complex plant problems is an everyday exercise that affects all projects and budgets, and it is very rarely introduced to students at any level. In fact, by the time young engineers are few years into working in industry, most have lost the engineering skills that they developed in school. We hope to encourage future engineers to retain those skills and learn the importance of plant problem-solving.

498677 Road Map for Embedding Ethics into ChE Undergraduate Curricula

Abstract:

The author will outline a number of methodologies and a suggested subject sequencing that can be effectively used to further develop an embedded ethics curricula. This curricula will consist of definitions, examples and suggestions on where to embed ethics into any existing ABET-approved, chemical engineering undergraduate course set. The author has lectured extensively at universities on ethics and a number of ethics related subjects.

504018 Views on Ethics in Undergraduate Education

Abstract:

Engineering programs are expected to graduate students who understand their professional ethical responsibilities. A studentâ€™s experience with ethics often begins with academic integrity. Research on student attitudes and behavior regarding academic integrity will be reviewed with suggestions to help faculty nurture a strong ethical foundation in their students.

495849 Ethical Reasoning in the Engineering Curriculum

Abstract:

Although the provision of ethics in engineering curricula has increased in recent years, the teaching of engineering ethics remains patchy and sporadic in a number of countries. Ethics in research, on the other hand, appears to be at a more mature stage. At the same time, the number of ethical codes, codes of practice and research concordats has flourished, both nationally and internationally. The questions remain: are those codes â€œliveâ€? Have those concordats resulted in an improvement of professional integrity? Is the teaching of ethics different and/or disconnected from ethics in research and in the workplace? Is ethics in chemical engineering different from ethics in other engineering disciplines? The paper analyses those questions in relation to an ever changing landscape, influenced by local and international drivers.

NCEES
advancing licensure for engineers and surveyors

EFFECTIVE AND EFFICIENT USE OF THE FUNDAMENTALS OF ENGINEERING EXAM FOR OUTCOMES ASSESSMENT

AICHE 2017 ANNUAL MEETING

David Whitman, PhD, PE
NCEES Past-President
University of Wyoming, Retired

NCEES
advancing licensure for engineers and surveyors

Overview

- Applicability of the FE exam for assessment
- ABET student outcomes that can be measured
- Computer-based FE exam format
- Various methods to use FE exam results for outcomes assessment
- Self-study examples and closing the loop
- Questions and answers

NCEES

Applicability of the FE Exam for Assessment

- It is a direct method of objective assessment with comparisons of institutional results against national results.
- Assessment does not use pass rates but how students perform on individual exam areas.
- Because there are over 55,000 FE examinees per year, it provides high reliability.
- The FE can be used as an assessment tool with a pool of all graduates or with a self-selecting pool.

NCEES

ABET Outcomes Assessment Possible with FE Exam

a) An ability to apply **knowledge of mathematics, science, and engineering**

b) An ability to design and conduct experiments, as well as to **analyze and interpret data**

c) An ability to **design a system, component, or process** to meet desired needs

NCEES

ABET Outcomes Assessment Possible with FE Exam (cont.)

(e) An ability to **identify, formulate, and solve engineering problems**

(f) An understanding of **professional and ethical responsibility**

(k) An ability to use the techniques, skills, and modern **engineering tools** necessary for engineering practice

NCEES

Computer-Based Testing (CBT) Format

- The FE became a CBT exam in January 2014
  - Shorter in time and fewer questions
  - Presented in a different format

NCEES

CBT Format (cont.)

- Length
  - The appointment time at test centers is 6 hours.
    - Tutorial–8 minutes
    - Nondisclosure agreement–2 minutes
    - Exam time–5 hours, 20 minutes with a 25-minute scheduled break after approximately 55 questions
    - Brief post-exam survey
  - Total of 110 questions

NCEES

CBT Format (cont.)

- Testing opportunities
  - Testing windows
    - January–March
    - April–June
    - July–September
    - October–December
- Test center locations
  - Pearson VUE test centers

NCEES

CBT Format (cont.)

- *FE Reference Handbook*
  - Provided electronically with the exam as a searchable PDF
  - Available for free download and for purchase as a hard copy at http://ncees.org/engineering/fe/
- *NCEES Examinee Guide*
- Computer-based practice exams

NCEES

CBT Format (cont.)

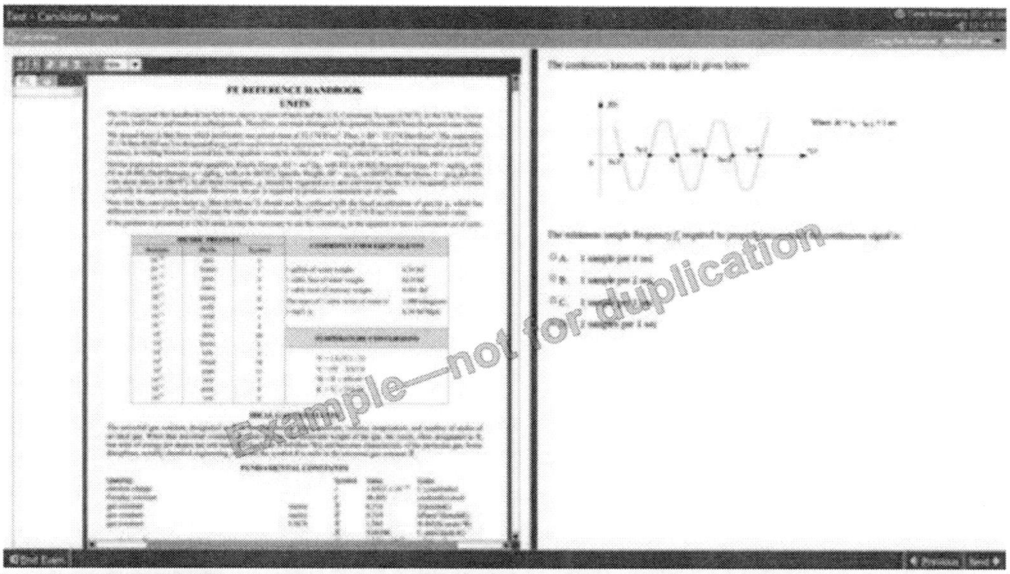

NCEES

CBT Format (cont.)

- Content of the exam
  - 7 free-standing discipline-specific exams
    - Chemical, Civil, Electrical and Computer, Environmental, Industrial and Systems, Mechanical, Other Disciplines
  - FE no longer has a common morning portion (breadth module)
    - Selected exam topics that were previously on the breadth module are now tested within each discipline-specific exam.

NCEES

FE Chemical Engineering Exam Specification

- Mathematics
- Probability and Statistics
- Engineering Sciences
- Computational Tools
- Materials Science
- Chemistry
- Fluid Mechanics/Dynamics
- Thermodynamics
- Material/Energy Balances
- Heat Transfer
- Mass Transfer and Separation
- Chemical Reaction Engineering
- Process Design and Economics
- Process Control
- Safety, Health, and Environment
- Ethics and Professional Practice

Additional details can be found at http://ncees.org/engineering/fe/

NCEES

CHE Example of Engineering Design

13. Process Design and Economics

A. Process flow diagrams and piping and instrumentation diagrams
B. Equipment selection (e.g., sizing and scale-up)
C. Cost estimation
D. Comparison of economic alternatives (e.g., net present value, discounted cash flow, rate of return, expected value and risk)
E. Process design and optimization (e.g., sustainability, efficiency, green engineering, inherently safer design, evaluation of specifications)

NCEES

So, what actual data are available, and what can you do with the data?

NCEES

Subject Matter Report

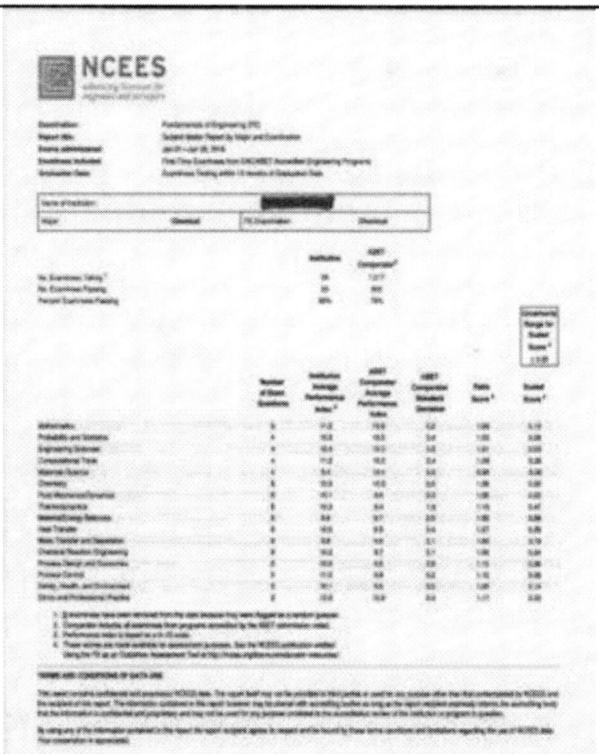

Subject Matter Report (cont.)

- Reports are generated twice a year.
  - In July for the January–June testing period (Spring)
  - In January for the July–December testing period (Fall)
- The report is specific to
  - An institution,
  - Students within an engineering degree program at that institution, and
  - The discipline-specific exam that those students completed.

NCEES

Subject Matter Report (cont.)

- Data are provided for all examinees testing within 12 months of graduation (either before or after graduating).
- Only first-time takers are included.
- Random guessers are removed from the report.
- National performance data, with standard deviation information, are also provided for the same degree program and same discipline-specific exam.

NCEES

Subject Matter Report (cont.)

- For each topic, the students' performance is given as a Performance Index on a scale of 0–15.
- The Performance Index is <u>indirectly</u> related to the average number of questions answered correctly.
- This is necessary because each examinee receives a different set of questions within each topic area.
  - Any one examinee's set of questions may be harder or easier than another examinee's set.

NCEES

Subject Matter Report (cont.)

- Getting the data
  - NCEES sends links to reports directly to an institution via email.
  - If you don't know, NCEES can tell you who receives your institution's reports.
  - Reports also include information on the specific institution's examinees who took the FE or PE exam more than 12 months after graduation.

NCEES

The specifics of using the FE exam for outcomes assessment

NCEES

Getting Started

- **Faculty** should be involved.
- Identify areas of strength.
- Acknowledge areas that are not emphasized.
- Set program-specific goals for each area.

NCEES

Suggested Analysis Techniques

- Choose your longitudinal time basis.
  - Performance from multiple examination windows
  - Academic year performance
- Choose your presentation method.
  - Ratio method
  - Scaled score method

NCEES

Ratio Method

- The ratio score is simply the ratio between the program's performance index (P.I.) in any topic area and the P.I. of the comparator performance.

  - Ratio score = Program P.I./Comparator P.I.

NCEES

Comparison of Ratios by Subject Area

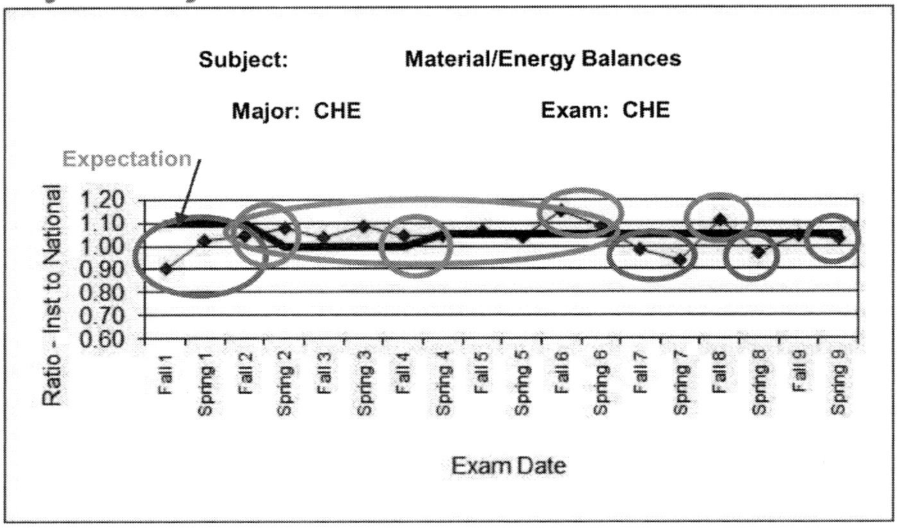

NCEES

Comparison of Ratios by Subject Area

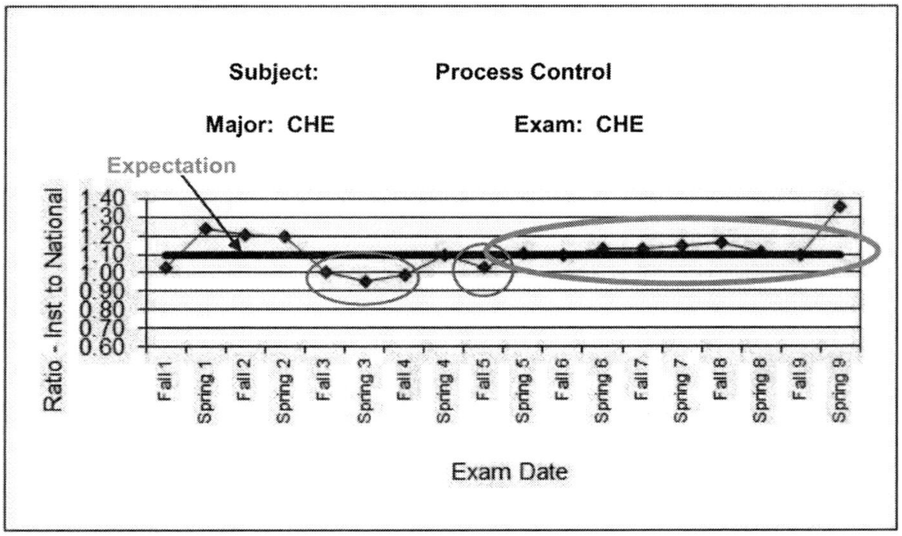

NCEES

Comparison of Ratios by Subject Area

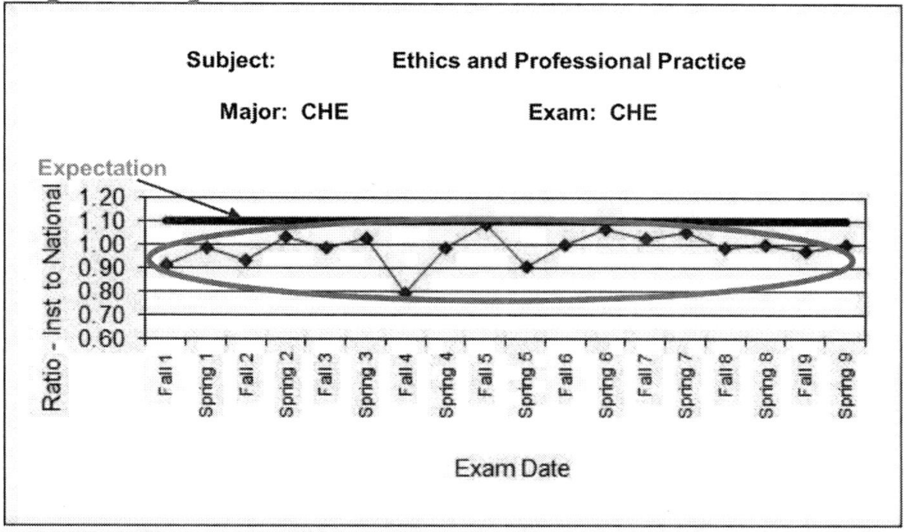

NCEES

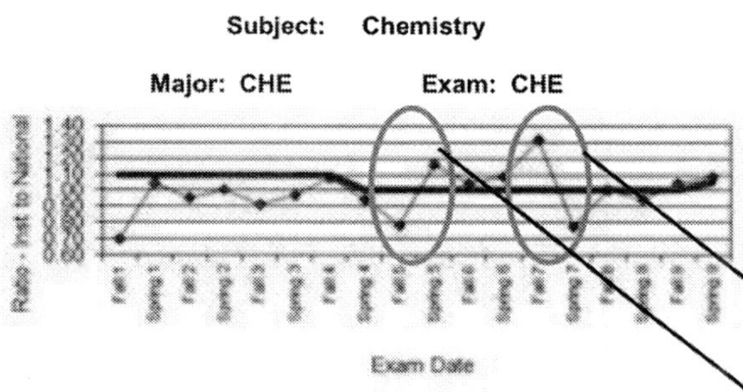

Effect of Academic Year Averaging

Causes a smoothing of the data—note the Fall/Spring 5 and Fall/Spring 7 results compared to the AY 5 and AY 7 academic year results.

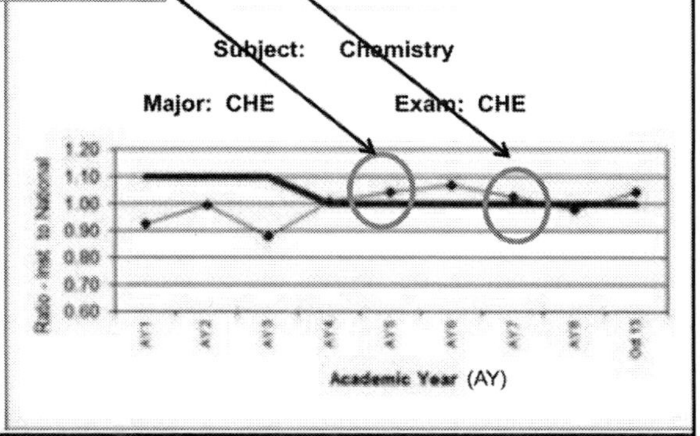

NCEES

Scaled Score Method

Define Scaled Score (S.S.)

$$S.S. = \frac{\text{PI for Univ X} - \text{PI comparator}}{\text{PI comparator standard deviation}} \pm \frac{1}{\sqrt{\text{\# of takers at Univ X}}}$$

NCEES

Comparison Using Scaled Score by Subject Area

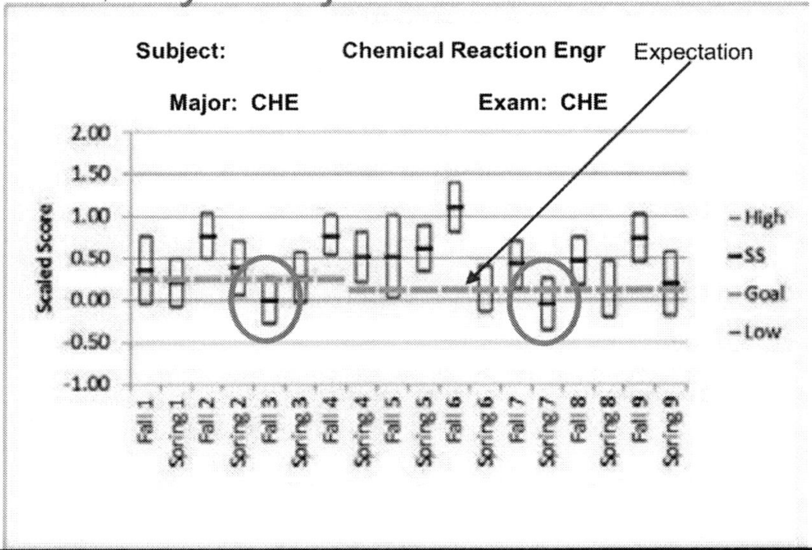

NCEES

Comparison Using Scaled Score by Subject Area

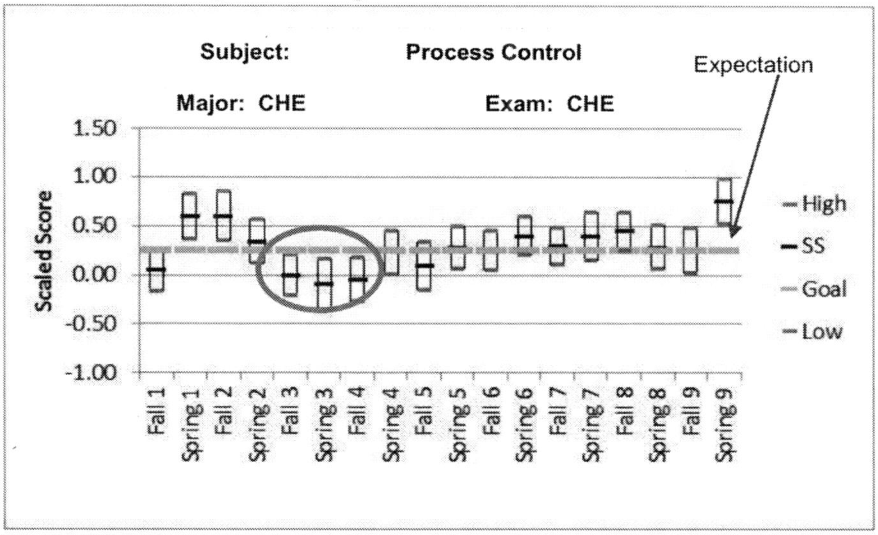

NCEES

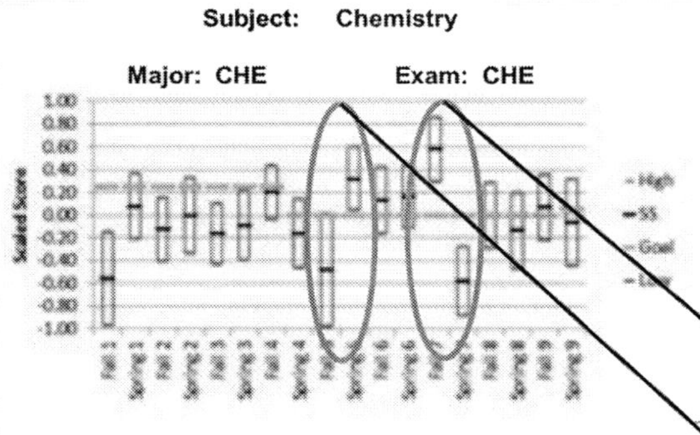

Academic Year Averaging

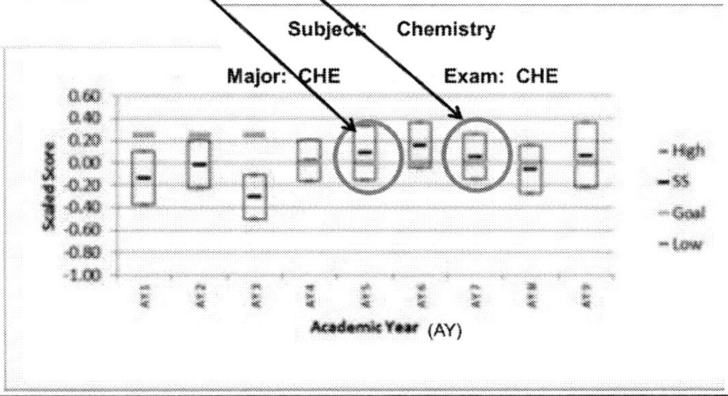

NCEES

Examples of Assessing Some of the ABET (a)–(k) Outcomes

NCEES

Ability to Measure ABET (a) Math and Science

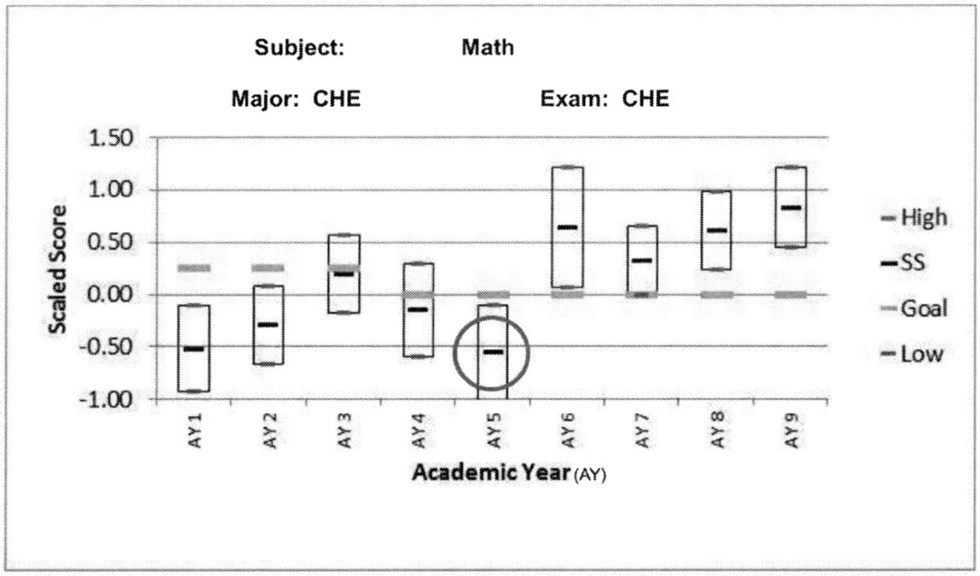

NCEES

Ability to Measure ABET (c) Design

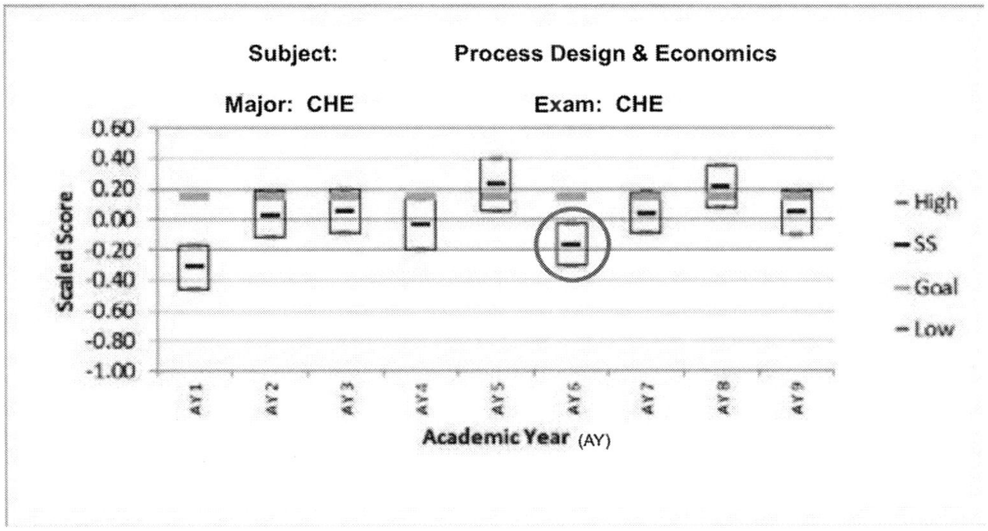

NCEES

Ability to Measure ABET (e) Solve Engineering Problems

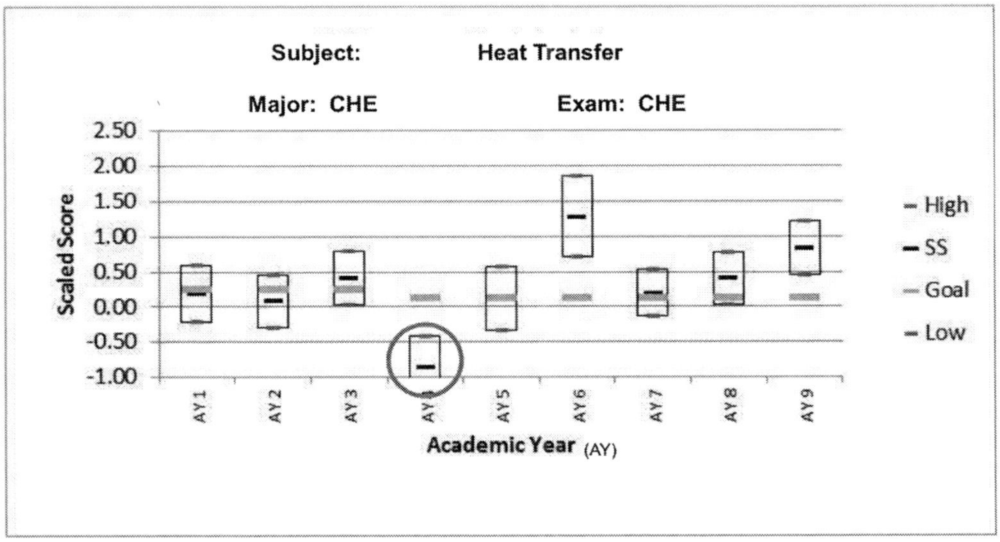

NCEES

Ability to Measure ABET (k) Engineering Tools

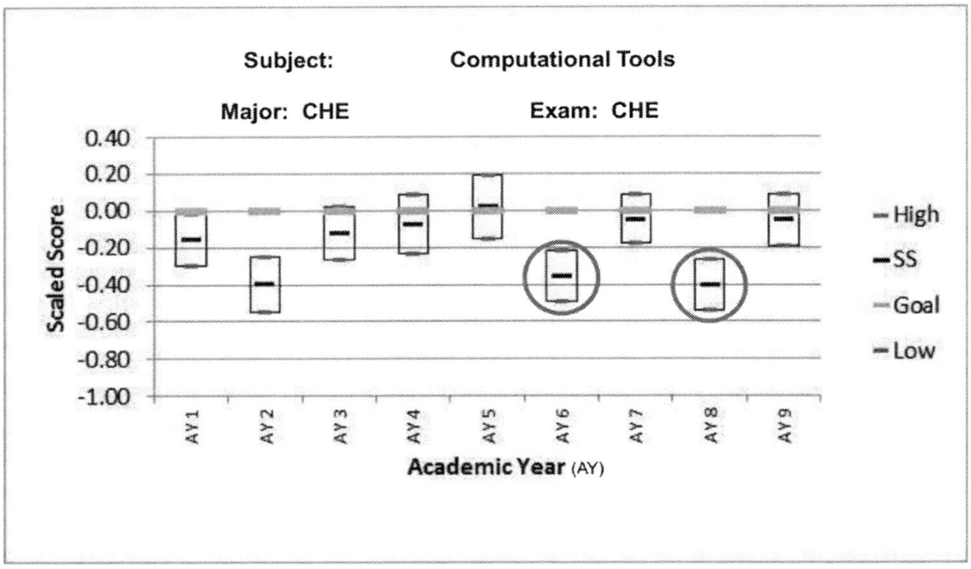

NCEES

Conclusions

- Discipline-specific assessment information can be gleaned from this exam.
- FE exam provides a direct, quantitative assessment technique.
- The Subject Matter Report provides comparative data.
- NCEES sends a link to the Subject Matter Report directly to your institution via email.

NCEES

Conclusions (cont.)

- The FE exam is one effective assessment tool to be used as part of your institution's full assessment package.

NCEES

Additional Resources

For more information on reports, email fereports@ncees.org

NCEES

Additional Resources

For further information, contact:

Cheryl Warren, Ph.D., P.E.
NCEES Exam Development Engineer
800-250-3196, ext. 5472
email: cwarren@ncees.org

NCEES

Questions?

NCEES

Dr. John Wagner P.E.

McKetta Department of Chemical &
Bioprocess Engineering
Trine University

My Background

- Passed the FE as a senior in ChE at University of Colorado – 1987
- Worked at Shell Oil– Deer Park TX
- Rice University for Ph. D. – 1994
- Accepted a Faculty Position at Trine University
- Passed PE in 1997
- Have been on the NCEES FE Committee since 1998

How we use the FE Exam fits into to the Assessment in ChE at Trine

Assessment Tool	Student Outcomes											
	a	b	c	d	e	f	g	h	i	j	k	l
FE Exam	X		X		X						X	X
CAE (Internal Comprehensive Exam)	X	X	X		X							X
Exit Interview /Survey	X	X	X	X	X	X	X	X	X	X	X	X
PAB - Professional Advisory Board						X	X			X		
Math Exam - Internal Comprehensive	X											
SAChE Student Program												X
Key Assessment Courses												
Process Measurements Lab												X
Numerical Methods											X	
Unit Operations Lab I		X										X
Unit Operations Lab II		X		X								
Kinetics	X											
Design I	X				X						X	X
Design II			X			X	X	X	X	X		

Color Code

Program Educational Objective 1	-	Technical Competency
Program Educational Objective 2	-	Problem Solving, Design, Communication and Team Skills
Program Educational Objective 3	-	Professionally Obligated

FE Exam is our only External, Objective Quantitative Assessment Tool

	Objective	Subjective
Internal	CAE Math Exam	Exit Interview Key Assessment Courses
External	FE Exam SAChE Certificate	PAB

Assessment Type	Student Outcomes											
	a	b	c	d	e	f	g	h	i	j	k	l
Objective, External	X		X		X						X	X
Objective, Internal	X	X	X		X							X
Subjective, External						X	X			X		
Subjective, Internal	X	X	X	X	X	X	X	X	X	X	X	X

FE Exam Results, 2010-2013

		2010		n=	9	2011		n=	12	2012		n=	18	2013		n=	18	4 yr Average	
AM Subject	#Q	Avg	Nat. Avg	Dev	Scaled	Avg	Nat. Avg	Dev	Scaled	Avg	Nat. Avg	Dev	Scaled	Avg	Nat. Avg	Dev	Scaled	Scaled	**AM Subject**
Mathematics	19	58	66	-8	-0.46	63	74	-11	-0.70	65	79	-14	-0.99	67	76	-9	-0.59	-0.717	Mathematics
Engineering Prob & Stats	8	61	65	-4	-0.20	69	69	0	0.00	47	54	-7	-0.33	62	61	1	0.05	-0.12	Engineering Prob & Stats
Chemistry	11	85	89	-4	-0.34	89	85	4	0.31	80	85	-5	-0.39	77	79	-2	-0.15	-0.158	Chemistry
Computers	8	71	76	-5	-0.29	69	76	-7	-0.40	68	77	-9	-0.60	81	82	-1	-0.06	-0.338	Computers
Ethics & Business Prac	8	77	76	1	0.05	80	73	7	0.43	80	85	-5	-0.33	82	80	2	0.13	0.036	Ethics & Business Prac
Engineering Econ	10	82	66	16	0.76	66	65	1	0.06	81	71	10	0.53	78	65	13	0.59	0.485	Engineering Econ
Eng Mechanics (Statics)	8	64	58	6	0.27	40	41	-1	-0.04	60	58	2	0.10	51	57	-6	-0.27	-0.02	Eng Mechanics (Statics)
Eng Mechanics (Dynamics)	5	78	68	10	0.42	48	62	-14	-0.54	36	42	-6	-0.23	40	42	-2	-0.08	-0.147	Eng Mechanics (Dynamics)
Strength of Materials	8	36	39	-3	-0.17	39	40	-1	-0.05	31	38	-7	-0.35	44	40	4	0.21	-0.081	Strength of Materials
Material Properties	8	67	65	2	0.11	61	60	1	0.05	51	56	-5	-0.25	69	62	7	0.35	0.061	Material Properties
Fluid Mechanics	8	69	71	-2	-0.09	69	62	7	0.28	63	66	-3	-0.16	69	72	-3	-0.15	-0.054	Fluid Mechanics
Electricity and Magnetism	11	56	55	1	0.05	50	55	-5	-0.23	45	61	-16	-0.77	44	50	-6	-0.31	-0.381	Electricity and Magnetism
Thermodynamics	8	51	54	-3	-0.13	58	64	-6	-0.30	76	76	0	0.00	82	73	9	0.45	0.058	Thermodynamics
Score	120	65	66	-0.3	-0.049	63	65	-2.3	-0.12	62	68	-5.8	-0.36	66	66	-0.3	-0.04	-0.159	Score
PM Subject																			**PM Subject**
Chemistry	6	57	58	-1	-0.05	85	77	8	0.40	56	70	-14	-0.65	78	74	4	0.24	-0.052	Chemistry
Mat/Energy Balances	9	75	72	3	0.16	72	73	-1	-0.05	70	68	2	0.09	82	73	9	0.45	0.186	Mat/Energy Balances
Chem Thermodynamics	6	63	62	1	0.05	64	58	6	0.30	65	65	0	0.00	64	59	5	0.21	0.139	Chem Thermodynamics
Fluid Dynamics	6	83	79	4	0.22	43	48	-5	-0.27	53	61	-8	-0.32	65	57	8	0.32	-0.023	Fluid Dynamics
Heat Transfer	6	74	74	0	0.00	71	73	-2	-0.10	68	80	-12	-0.60	73	71	2	0.09	-0.181	Heat Transfer
Mass Transfer	6	46	47	-1	-0.05	69	60	9	0.42	61	69	-8	-0.37	50	50	0	0.00	-0.036	Mass Transfer
Chem Reaction Engr	6	65	68	-3	-0.13	81	77	4	0.17	62	69	-7	-0.30	68	71	-3	-0.14	-0.123	Chem Reaction Engr
Proc Dsgn Econ Opt	6	72	70	2	0.10	85	75	10	0.50	73	72	1	0.05	89	76	13	0.60	0.326	Proc Dsgn Econ Opt
Computer Usage and Chem	3	78	67	11	0.41	61	67	-6	-0.23	69	75	-6	-0.23	89	68	21	0.70	0.168	Computer Usage and Chem Eng
Process Control	3	41	35	6	0.23	92	68	24	0.90	80	68	12	0.40	89	75	14	0.53	0.517	Process Control
Safety Health and Environ	3	85	81	4	0.17	50	53	-3	-0.13	54	61	-7	-0.21	76	75	1	0.04	-0.053	Safety Health and Environ
		67	66	1.7	0.079	71	67	3.6	0.161	64	69	-4.6	-0.206	74	67.65	6.1	0.264	0.064	
Score	60																		

Trine was not a testing site starting 2014. We stopped requiring the FE and our participation dropped to only 3 or 4 students/year

Students graduating in 2019 will once again be required to take the FE

NC STATE UNIVERSITY

Dr. Lisa G. Bullard, PE

Chemical and Biomolecular Engineering
North Carolina State University

NC STATE UNIVERSITY

My Background

- Passed the FE as a senior at NC State
- PhD → Eastman Chemical Company (Engineering & Construction Division)
- Took/passed the PE after two years of industry experience (plus 2 from PhD)
  - "Unwritten rules" about promotion
  - Changes in state regulations
  - Company support
- Faculty member at NC State

NC STATE UNIVERSITY

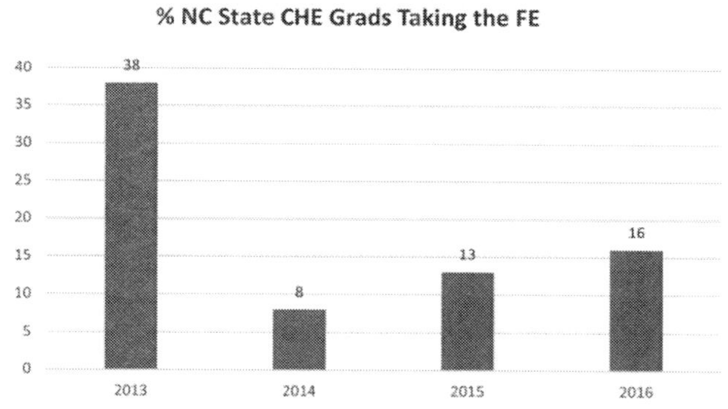

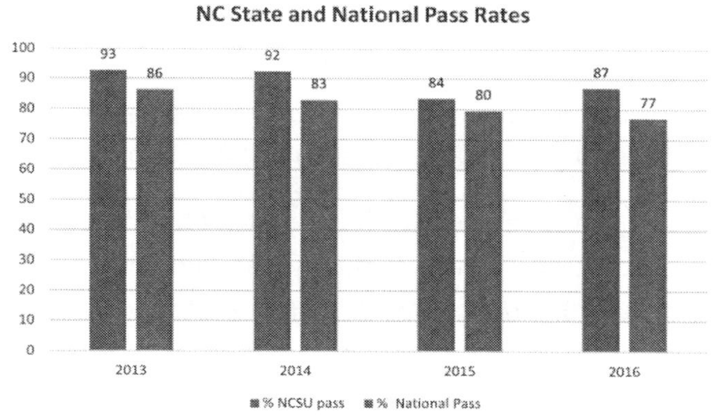

NC STATE UNIVERSITY

Issues/Questions

- Do companies hiring our students expect/promote the PE?
  - Industry
  - Functional role
- When is the best time for students to take the exam – in school or after graduation?
- Increased cost to students
- Resources for test preparation

<u>Using a Structured Approach to Efficiently Use the Brains of Other to Make Problem Solving More Productive</u>

Jack Hipple, Principal

TRIZ and Engineering Training Services LLC

Tampa, FL

813-994-9999

www.innovation-triz.com

In our attempts to problem solve and produce novel and breakthrough ideas, we frequently rely on psychological stimulation techniques and processes. These are effective for simple problems but only make use of the knowledge base of the participants. They may be "fun" techniques, but they do not take advantage of the global knowledge base available to us.

Around 1990, a powerful "left brained" problem solving approach, long practiced in the former Soviet Union, came to the US and is now used by many Fortune 500 companies to deal with their most difficult engineering and design challenges. This process and technology is known as "TRIZ", a Russian acronym for "Theory of Inventive Problem Solving". It uses the data base of breakthrough patents, and the methods and principles used in them, to provide a "left brained" algorithm useful for problem solving. It, in effect, brings the other inventors of the world into your problem solving session and avoids the barriers to problem solving such as ego, acronyms, and fancy language that reinforces our belief that the problem we have is special and unique.

The TRIZ problem solving algorithm, sometimes known as ARIZ (Algorithm for Inventive Problem Solving), contains several key aspects. First is the very basic concept of parallel universes of invention and using the most basic of terminology to search for potential solutions in parallel universes not previously seen, due to the use of company or industry specific language. For example a search for information about fraud would produce far fewer ideas than one based on the term, "substitution of one thing or another". Second, we envision and describe, in several different ways, the Ideal Final Result—without worrying about how we will get there. This is quite different than the normal engineering approach of looking for a 10-20% improvement. Then we look for available resources that our system and its sub-system and super-system might have available to achieve this "IFR". The TRIZ list of resources is far broader than the traditional chemical engineering look at material resources. We look at time (before, during, after), fields, field conversion opportunities, and the negative aspects of our system. This latter aspect turns out to be a surprisingly powerful tool for problem solving.

These basic steps are often sufficient to solve a problem, but often there are contradictions that need resolved. The resolution of contradictions is the key to most of the top 5% of unique patents, and analysis of these patents shows a limited number of inventive principles that are repeatedly used in these patents. TRIZ organizes these inventive principles in a useful and retrievable way, allowing their use in other problems with major contradictions that need resolved.

This workshop will overview this process and show by many examples where it has been applied.

487650 AIChE Engage: Your Next Stop for Brainstorming in the Process of Problem Solving or Innovating

Abstract:

Being an AIChE member provides you the opportunity to easily connect and build relationships with other chemical engineers in the world even if you donâ€™t have the time or ability to meet them in person. The newest benefit of membership, AIChE Engage, makes doing this even easier and more interactive than before. AIChE Engage will be your next stop for brainstorming in the process of problem solving or innovating.Â You can get two benefits through AIChE Engage:

Â 1.There is an extensive search engine which enables you to find other members by their names, companies, geographic locations, job functions and industries. There are also focused AIChE member directories of Fellows, Professionals, Young Professionals, Graduate Students, Undergraduate Students and members of the divisions, forums and committees. After youâ€™ve found the members you are looking for, you can send them an invitation to communicate via Engage or through another channel.

2.Perhaps the new feature of Engage that you will find most exciting is Engage Discussion Central. Professional members now have the opportunity to post and participate in discussions with colleagues all over the world on topics related to chemical engineering. Seeing the discussions is easy as they are sent directly to your e-mail address in a daily digest.

Therefore, AIChE Engage has become and will be a mobile-friendly platform for the global collaboration with your fellow members to get academic and professional accomplishments, career achievements and other distinguishing industrial breakthrough. This presentation will provide you the introduction on how to apply the methodology of TRIZ problem solving and strategic planning process with AIChE Engage.

Introduction to the Fundamentals of Project Management

Abstract ID: 481742
Paper # 330a
Dr. Eldon R. Larsen
Marshall University

In this paper we discuss several important concepts which will be helpful throughout the rest of the tutorial. First, we discuss the very important concept of *principles*. Then, we will seek to understand what we really mean when we say we have a project—in other words, "What is a project?" Finally, we will introduce the most important keys to successful project management, which will really be the foundation for much of the rest of the tutorial.

Principles: The Foundation

To begin, we need to understand the concept of what a *principle* is and how we can use that concept to strengthen our ability to manage projects. In our discussion of principles, the following definitions help us to understand some important ideas applicable to project management.

Tools. Tools are instruments, concepts, procedures, or techniques used to help us accomplish work, either physical or mental. A Gantt chart, to be discussed in a later paper, would be an example of a project management tool. Tools are not our masters; they are intended to be our helpers or servants. Tools can be applied in many ways and they can be misapplied in many ways. Tools need to be maintained properly in order for them to continue to help us effectively. We use tools in applications of rules, policies, and principles.

Rules. Rules are prescribed guides, procedures, or obligations for conduct or action and are sometimes binding by regulation or legal precept. An example project management rule might be that we don't have employees work overtime for more than two weeks straight without a two-day rest break before the next overtime period. Good rules are based upon some kind of principle(s), while bad rules probably have little or no foundation in principles. Rules occasionally need to be changed, especially as times and circumstances change. In our example overtime rule, a true emergency might require an exception.

Policies. Policies are very much like rules. In fact, they are often interchangeable in definition. They often consist of management procedures based on material interests. They might involve a definite course of action selected from among alternatives and in light of given conditions to guide and determine present and future decisions. An example policy might be that we will give an employee one day off if he or she needs to attend the funeral of an immediate family member. Like rules, policies do change and often need to change with time and with new circumstances. There might need to be exceptions to the example funeral policy, if the funeral was more than a given distance from the workplace. Like rules, good policies are based upon some kind of correct principle(s), while bad ones are not.

Theories. Theories tend to be derived from various analyses of facts and their relation to one another. They usually are plausible or scientifically acceptable general concepts offered to explain phenomena. They usually are based on observations and are: hypotheses assumed for the sake of argument or investigation; suppositions or conjectures; or abstract thoughts or speculation. Because theories are not facts, but rather are attempts to explain facts or observations, theories often change or need to change, especially when new facts are obtained or when new observations occur. It is not uncommon to read in the newspaper of some theory that has changed due to new information which someone has obtained. In project management, we occasionally encounter various management theories. Some are well-based on extensive evidence, while others are less-well based.

Values. Values are concepts or perceptions which we highly rate or highly regard—something we prize. They may or may not be correct perceptions, but to each of us they are important in our decisions and actions. Religions, philosophies, personal ideals, organizational ideals, and some morals might fall under the category of values. For example, it might be that a person highly values self-reliance to the point that he or she feels it is important to become wealthy before truly being self-reliant. His or her value is that he or she needs to become wealthy. However, it is not unusual for people and organizations to change their values over time, even if only a little, as a result of experience, knowledge, and choices they make in life. In our example value of the need to become wealthy, the same person may encounter life in such a way to cause him or her to re-evaluate the desire to become wealthy and replace it with a desire to sacrifice of higher causes. Thus, one's values do occasionally change.

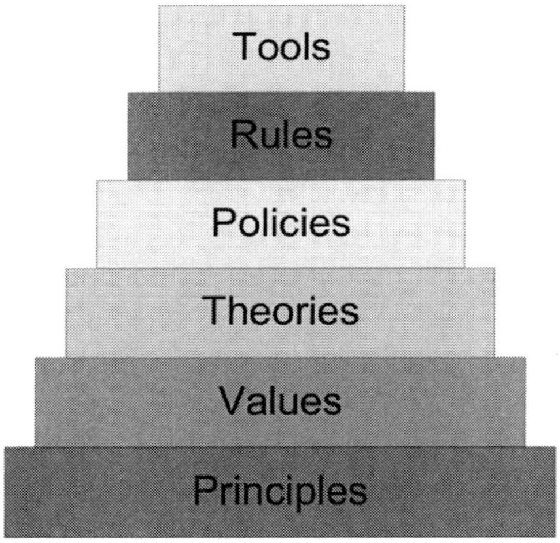

FIGURE 1Good Project Management Should Be Founded Upon Principles

Principles. Principles are fundamental or foundational unchanging truths or natural laws that cannot be broken. Principles are different from all the other concepts and definitions above in that *they do not change* with time or circumstance and they are true for everyone; regardless of background, religion, or philosophy; regardless of whether we choose to ignore them. They are comprehensive and they are self-evident. A few examples of principles are given in Table 1.

Principles are not the same as the *application* of principles. Applications of principles might change a little or a great deal as time or circumstances change.

For example, it is important that we love people—this is a principle. In some cases this might mean giving something to someone in need, like giving food to a child who cannot provide for himself. In other cases, loving an adult might be applied by *not* giving him something he needs, particularly if it is important for that adult to learn how to obtain that thing through his own skills and initiative, so that he can grow and become more self-reliant—note that the importance of growth is also a principle. In the examples of the child and of the adult, love is evident, but the practice or applications of the principle of love are very different.

Table 1: Examples of Principles
The need to love and care about other people
The necessity of integrity
The importance of honesty
The importance of service to other people
The importance of human bonding
The need for patience
The need for good communication
The need to be loved
The need for individual growth
The need for harmony with one's conscience

We must note that while we can try to ignore principles, they never go away, they are always there. We can try to say that they are not true, but that does not change that they are true; and when we try to break principles, we must suffer the natural consequences, like we do when we try to break natural laws.

Because principles are so very fundamental, unchanging, and universally applicable, it is vital that we learn them, understand them, and learn to apply them correctly. This is true in project management and in any other aspect of life.

Throughout the use of this text, you are encouraged to focus special attention on learning the *principles* of project management. Although we will be considering a number of procedures, tools, and policies or rules, all these can all change over time—they will change as a function of circumstances. The principles however *will not* change.

If the reader deeply understands the principles of project management, which will never change, then he or she will have a firm foundation upon which to build various different applications of those principles, according to the needs of the situation. Then, we can also understand how to make better adjustments or exceptions to rules and policies as needed.

What is a Project?

Before we go any farther, we really ought to understand what a project is. This might seem trivial, but it is not. In fact many individuals and teams have started out to accomplish a "project" only to find later they really did not have any clear idea of what was to be accomplished and when.

There are many people and organizations who have attempted to develop a definition for a *project*. Because it covers the basics of all projects, we will use the following definition: A project has two main characteristics: (1) it is temporary, or time limited; and when completed (2) a project provides a unique product, service, or result, or in other words the project has one or more specific objectives. See Table 2.

PMI® (The Project Management Institute®) says it this way: "A project is a temporary endeavor undertaken to create a unique product, service, or result." (PMBOK3, p. 17)

Table 2: The Essential Characteristics of a Project

1. Temporary; or in other words the project is time limited
2. Provides a unique product, service, or result; or in other words the project has one or more specific objectives.

Temporary. *Temporary* means that the project is time limited—it has a beginning and a end. It is the project that is temporary, not the product, service, or result—these normally last much longer than the project. The project cannot go on forever, even if some people treat it that way. Some projects might end in a matter of hours or days, while others may take many months or years, but each project must end; and it is important that the project stakeholders have a relatively clear expectation for when the project is to end.

We say "relatively clear expectation" because, while for some projects we can give a nearly exact end date or time, there are projects for which we can only give a very approximate range for the end time. This applies particularly to new technology projects that are highly uncertain and have high risks involved, like many R&D (Research and Development) projects.

For example, we might say that project A, which is designed to find a way to prevent wide-spread bird-flu infections, must finish within the next three to four years. In this case, although the end-date is not exact, we do have a clear expectation that we cannot go over four years, and that ideally we should try to finish within three years.

The bottom line is this: whatever the situation, there should be a clear expectation as to the time frame for the project to start and end, along with any tolerances in those start and end times.

Unique. By the time a project is completed, a that project should provide one or more unique products, services, or results—it should have accomplished one or more specific objectives. *Unique* does not mean that similar projects have not been done before, but rather that every project is unique in some way.

The project must have specific objectives to be accomplished. Without specific objectives, the project team's work will be inefficient and ineffective. The less specific the objectives are, the greater the likelihood that there will be much wasted effort, increased rework, and missed opportunities. In addition, if the objectives are not clear, the project stakeholders will have a difficult time agreeing on when the project has been completed and whether or not the project has provided the product, service, or result each stakeholder anticipated.

Non-project Work. Work that falls outside of project work is referred to as being *operational work* or *maintenance work*. These types of work tend to be ongoing and repetitive. At home for example, the work of washing the laundry, cooking the meals, washing the dishes, and cleaning the house is very repetitive and ongoing—it seems it never ends! Although people might say that cleaning the house is a project, for the purposes of project management we normally would not consider that type of work as *project work*, but rather we consider it as *operational work* or alternatively *maintenance work*, whichever you prefer—it is not unique and really is not temporary.

What is a Project Manager?

The obvious answer to the question, "What is a project manager?" is that a project manager is someone who manages projects. Perhaps we should re-phrase the question into the following: "What characteristics make up a good project manager?" Table 3 gives a list of characteristics that the author has compiled over the years which cover most of the aspects of what is needed to be a good project manager. The list is large and yet probably lacking. Good project managers need to have many talents and many excellent character traits if they are to be successful project managers and project leaders. Many of the characteristics in Table 3 are covered in detail later in this text.

Table 3: Characteristics of a Good Project Manager

Maintains high integrity

Excellent communicator and listener

Trustworthy

Honest and ethical

Friendly and courteous

Respectful *and* respectable

Organized

Teacher and builder of people

Patient

Accountable and asks for accountability

Cares about people

Flexible when needed; not flexible when shouldn't be

Motivator

Professional in interactions with others

Good decision maker

Planner, with other stakeholders

Good manager of time

Good record keeper

Has good technical and managerial credibility

Familiar with the organization's culture, policies, and procedures

Good diplomatic skills

Receptive to suggestions

Skilled in developing team solutions and facilitates group participation

Doesn't ask others to do what he or she is not willing to do

Leads rather than pushes

Values the differences in others and their perspectives

Seeks synergy and win/win solutions

Helps others set goals in support of the project and organizational objectives

Keys to Successful Project Management: The Vital Three

Let's suppose that you are nearing the end of this project management text. You meet a friend in the elevator who says, "I understand you attended a tutorial about project management. What have you learned?" You have until you exit the elevator, about 20 seconds, to tell your friend what you have learned. What would you say? Well you don't know right now because you haven't completed this presentation. So, I am going to tell you what I hope you will say. You see, I am going to tell you right now what I consider to be the three most important concepts in this tutorial. In the elevator, you would likely only have time to tell your friend the three most important concepts anyway. By me telling you now what I want you to say, I am hoping that you will pay special attention to these items throughout the course of your study. The three most important keys to successful project management are given in Table 4.

Table 4: The Vital Three Keys to Successful Project Management

***First**, the importance of **people** in project management!*
***Second**, the importance of excellent definition of **project objectives**!*
***Third**, the importance of excellent **planning** and **re-planning**!*

There are many concepts, processes, and principles important in project management and they are all important, but I consider these three to be the most important keys. If anyone of these is missing or weak, the entire project will suffer tremendously. Sometimes I refer to these as the *vital three* because they are so preeminently important. You can often succeed by neglecting some other aspects of project management, but you cannot ignore any one of the vital three—they are essential.

References

Covey, Steven R., *The Seven Habits of Highly Effective People*.

PMI. *A Guide to the Project Management Body of Knowledge (PMBOK® Guide)*, fourth edition, The Project Management Institute, Newtown Square, Pennsylvania, 2008.

The Importance of People in Project Management

Abstract ID: 481743
Paper # 330b

Dr. Eldon R. Larsen
Marshall University

Of the many important aspects of project management, the most important key to successful project management is people. When we speak of project management, we often think of Gantt charts, time lines, budgets, resource constraints, quality challenges, performance measures, and so on. While these are important, they are neither meaningful nor effective without people. This paper presents and discusses several fundamental concepts regarding the importance of people in project management.

Introduction

In 1993 when I first became deeply involved with formally studying and applying the principles and tools of project management, I made an appointment to visit a man who was well respected at for his project management expertise. He was Charles Brod, the Chair of the UCC (Union Carbide Corporation) Project Management Council. This council was made up of the most experienced project managers in the corporation.

I had been asked by UCC to lead a team to develop and implement project management principles, techniques, and tools specifically for research and development, an area often not accustomed to the rigors of formal project management. I had been on many research and development project teams and product or process commercialization teams, once serving on twelve different teams at the same time. However, I had little formal project management training at the time. I knew I could use Charles's insights for this effort I had been assigned.

I did some preparation and wrote questions to ask Charles, so as to make the most of our visit. As I entered his office at the appointed time, and before we sat down, I said, "Charles, I know that people are important in project management..." Before I could finish my sentence he pointed his finger at me and said, "You stop right there! People aren't just important in project management, people are project management!" (Brod, 1993) I just stood there stunned for a few seconds as my brain rapidly processed what Charles had said: "People are project management!"

As my mind raced through my many experiences in the many projects I had worked on, I knew almost instantly that what Charles had just said was not only true, but was the verbalization of one of the most important principles of project management. In comparison to that statement, little else mattered in that interview.

Principle:

People are project management!

People Are Project Management

Jack Gido and James Clements (Gido and Clements, 1999) say it this way: "It is the people—not the procedures and techniques—that are critical to accomplishing the project objective. Procedures and techniques are merely tools to help people do their jobs."

As people progress in their careers, many become project managers. These people are often highly trained technically and tend to be very analytical—this helps them to be good technical problem solvers. However, technical professionals often have had little formal training which will enable them to effectively lead and manage people. Thus, they could be weak people-problem solvers. Yet, this is exactly what they are being expected to do as they move into project management or other management positions.

People in new project management positions might well ask themselves the following questions: "What do seasoned project managers wish they had known on their first project management assignment?" "What are the responsibilities?" "What are the people interactions that take place?" Following are some of the things new project managers would have liked to know, based on research by J. Pinto and O. Kharbanda (Pinto and Kharbanda, 1995).

- Realizing that conflicts occur as a natural part of team development,
- Knowing who are the project stakeholders and what are their agendas,
- Realizing that organizations are political,
- Understanding that project management is leader-intensive and you must be flexible,
- Realizing that customer satisfaction is essential for project success,
- Understanding that you must be a motivator, coach, peacemaker, and conflict manager,
- Understanding that your team develops attitudes based on your emotions,
- Knowing that you always need to ask "What if…"; don't be comfortable with the status quo, and
- Learning that above all you must plan, plan, plan!

All of the above learnings involve *people interactions*!

Project managers have responsibilities among other things to (1) define and clarify objectives, (2) lead and manage, (3) plan and re-plan, (4) organize, and (5) control and influence. Again, all of these responsibilities involve people interactions.

The following sections of this paper discuss and make suggestions for some of the multiple interactions among people involved with projects. The interactions discussed are

- Stakeholders and stakeholder management
- Project manager interacting with project team members,
- Project manager interacting with senior managers,
- Project manager interacting with line mangers,
- Project manager interacting with the other employees within his or her company,
- Project manager interacting with the general public outside of his or her company,
- Project manager and team interacting with the customer, and
- Project manager and team interacting with contractors and other internal or external service providers.

Stakeholders and Stakeholder Management

All projects have project stakeholders. Examples of potential stakeholders are the customer, the project sponsor, senior managers, the project manager, the project team, organizational line managers, contractors and subcontractors, service and trades personnel, landlords and/or leaseholders, suppliers, recipients of products and technology, government entities and regulators, and possibly the surrounding populace.

A wise project manager will do what is needed to identify all significant stakeholders. The project manager will recognize that some stakeholders that might not seem significant at the present, but they might become significant and so will need to be monitored. One major mistake many project managers make is not to do a careful analysis of all stakeholders.

Stakeholders can be either positive stakeholders or negative stakeholders. Generally, positive stakeholders will benefit from the project being a success and will want it to succeed. Negative stakeholders will likely not benefit from the project's success, and so will want it to fail. Some stakeholders that currently are positive might change and become negative, and vice versa.

You might need the full support and cooperation of some stakeholders in order for the project to succeed. Therefore, you should always ask yourself the following questions: "Who will benefit from this project? Who might be hurt by this project? Who might otherwise be impacted by this project? Who will be key in helping this project to succeed? Who else might I have forgotten about?"

Because of the many concerns, needs, and expectations of stakeholders, it is essential that the project manager conduct an effort to identify the stakeholders, their expectations, their potential impact on the project, how the project impacts them, what is needed from them, and so forth. To do this, the project manager develops a *Stakeholder Register*.

Table 2.1 shows an example of a possible stakeholder register. Here we identify the stakeholders, their positions in the internal or external organizations, their power and authority, their location, how to contact them, whether they are positive or negative stakeholders, their relationship to the project, what influences they might have upon and for the project, their needs—must have, their wants—like to have, and what concerns or fears they might have about the project. We also very clearly identify their requirements for the project and their expectations beyond their requirements. We clearly identify our strategies for managing their requirements and expectations so that changes are clearly analyzed, controlled, approved or not, and communicated. Finally, we clearly define communication needs for each stakeholder.

The rest of this paper discusses some specific needs and helps in dealing with specific types of stakeholders.

Project Manager Interacting with Project Team Members

Managers are not necessarily the same as leaders. A manager has assigned people that he or she is to manage. Leaders are different. There are no leaders without followers, and people choose to follow those they consider to be their leaders. Leadership and management are all about working with people.

Much like a music conductor (Gido and Clements, 1999) leads an orchestra to play a concert, the project manager leads the project team to accomplish the project objectives. The project team, like an orchestra, must be unified and in tune (aligned), as they follow their leader. For there to be harmony, the team must want to follow the leader—this is a vital people interaction principle. The leader and team must care enough about each other to make this happen.

Project managers and the project team members interact in many ways. They do this as they define and clarify project objectives; plan and re-plan; make and accept task assignments; delegate and expect accountability; manage and resolve conflict; reward and discipline; generally communicate; give performance feedback; motivate and inspire; and develop personnel. Following are some suggestions for project managers to help them to achieve success in these interactions with team members.

- When practical, involve the project team in refining the project objectives with the customer;
- Assure that the project objectives, constraints, and deliverables are clear and that they are understood by all team members;
- Always involve the project team in project planning, then...
- Seek commitment from team members;
- When challenges arise, genuinely ask for input from team members;

Table 2-1 Example Stakeholder Register

STAKEHOLDER REGISTER

Stakeholder Characteristics

Stakeholder Name	Position	Power & Authority	Location	Contact Information	Category: Pos./Neg.

Stakeholder Relations and Needs, Wants, and Fears

Stakeholder Name	Relation to the project	Potential Influences	Needs	Wants	Fears

Stakeholder Requirements, Expectations, Strategies, and Communications

Stakeholder Name	Stakeholder requirements statements based on needs, wants, fears, and other information	Stakeholder additional expectations with respect to and for the project	Strategies and issues or concerns for managing the requirements and expectations	How, when, and what to communicate with this stakeholder

- Delegate as much as you can to team members, according to each individual's readiness to accept the added responsibility;
- Build and develop your followers; give them growth opportunities, but don't overwhelm them;
- Whenever feasible, minimize unnecessary bureaucracy;
- Learn what motivates and inspires each team member;
- Ask for accountability, while helping team members to get the authority and other resources they need in order to be able to be accountable;
- Work with line management to assure that team members have up-to-date tools;
- Reward or discipline team members sooner, rather than later;
- Make project meetings and communications valuable—don't waste people's time; (Larsen, 2001)
- When conflict arises, manage it quickly and seek win/win resolutions;
- Make a project communications plan and use it; and
- Keep accurate and up-to-date project records—this helps to minimize miscommunications and unnecessary rework.

Project Manager Interacting with Senior Managers

There are many reasons that a project manager needs to interact with senior managers and to nurture these relationships. Senior managers

- Assure business goals are being achieved,
- Often act as the interface between the customer and key project personnel,
- Provide essential support and guidance for projects, and
- Often act as project sponsors or champions.

It is absolutely essential that effective interactions occur between the project manager and senior managers. If this interaction with senior management is not effective, support for the project could wane, misunderstandings will almost surely occur, project objectives and priorities will either be unclear or insufficiently acted upon, and the project team could easily feel itself drifting without a good sense of direction or support. Some of the interactions that often occur between the project manager and senior managers have the following purposes.

- Define project objectives and deliverables,
- Clarify and align business and technical objectives,
- Clarify company policies and procedures,
- Clarify project scope and project risks,
- Clarify critical success factors,

- Clarify key project priorities or drivers and tolerances in the triple constraints of time, resources, and performance, (Larsen, 2002) and
- Finalize and approve the project charter.

In order for the above interactions between the project manager and senior managers to be effective, consider the following suggestions.

- Be honest, maintain integrity, be respectful;
- Don't be afraid to ask for or give clarification or more information as often as is needed;
- Give and ask for the accurate, whole picture, not just the good news;
- Be concise in communications, but be clear;
- Do all you can to prevent surprises;
- Be sure to thoroughly discuss and understand all important project risks, tolerances, and priorities;
- When discussing concerns, focus on issues rather than people;
- Wisely use senior management to help with conflict and risk management when their help is truly essential; and
- Never backbite your senior managers—it won't help and will likely make things worse.

Project Manager Interacting with Line Managers

Line managers frequently play a vital role in contributing to project success or failure. In many organizations it is the line managers who select whom in their groups to assign to a given project. In many organizations it is also the line manager who actually pays the employees' salaries on a team. This can cause a definite allegiance preference for the line manager over the project manager when it comes to deciding which boss to follow when getting mixed direction signals. In organizations where line managers appoint team members, these managers

- Usually provide the people to work on the project;
- Assure that their people have proper training;
- Monitor the workload of their people;
- Often take care of the administrative needs of their people (payroll, safety certification, annual reviews); and
- Are usually much more technically expert than is the project manager.

From the above list it should be obvious that line managers can be vital to your success! Some suggestions for working with line managers are

- In the very early stages of the project, begin communicating with the line managers of people who might be selected to be on the project team;
- Assure that the project objectives, deliverables, priorities, drivers, and tolerances are clear to each line manger who will be providing resources or services;
- Develop and nurture a close working relationship—you need line managers' support, advice, wisdom, and buy-in;
- Always keep line managers informed of the project progress, concerns, challenges, and needs;
- Keep line managers aware of personnel achievements, performance issues, or needs pertinent to their areas of responsibilities;
- Listen sincerely to the concerns and advice of line managers—if you don't, you will pay for it; and above all
- Communicate well and regularly. (Larsen, 2003)

Project Manager Interacting with the Company and General Public

Depending on the project, a project manager's interaction with the company as a whole or with the general public can range from non-existent to frequent and crucial. Corporate or public perception of you and your project can have the potential to make or break your project—if this applies to you, nurture appropriate interpersonal relations with all the care you can give. Following are suggestions.

- Learn what major concerns the company or the public is likely to have and, as appropriate, be fully prepared to address them honestly and openly. If you can't be open because of secrecy agreements or other legal constraints, make that clear to those who are concerned.
- Always be prepared! Know exactly what is happening on the project. Then plan and practice how to communicate with those concerned. Make sure senior management, the customer, and the project team know what you will present and that they have given you their suggestions and support.
- Try to put yourself in the shoes of the people with whom you need to communicate. Seek to understand them (Covey, 1989).
- Always be ethical in your behavior and your words.
- Do not lose your cool! It takes many good interactions with people to make up for one bad one.

Project Manager Interacting with the Customer

The interaction the project manager has with the customer is perhaps the most important interaction of all, and could have the greatest impact on customer satisfaction—the real measure of project success. The customer, better than all others, usually knows most what he or she wants at the end of the project. Customer perceptions play a strong project-success role. Effective people-interaction is necessary to transfer customer expectations to the project manager and team.

Following are suggestions to help in this interaction.

- Unless the nature of the project prevents this, the project manager and team need close interaction with the customer as early in the project as possible—this is when exists the greatest potential to affect the final project outcome. Make this happen if possible.
- Maintain open channels of communication throughout the entire project.
- Early interactions between the project manager, team, and customer should include developing a very clear understanding of the
  - ✓ Project objectives,
  - ✓ Project priorities, triple constraint drivers, and tolerances,
  - ✓ Project's critical success factors,
  - ✓ Project's scope and deliverables,
  - ✓ Project background and justification, and
  - ✓ Any key concerns and assumptions.

Project Manager Interacting with Contractors and Other Internal or External Service Providers

Outsourcing of project work has increased greatly over the past decade. Flannes and Levin (2001) point out that those temporary workers sometimes "contribute a greater depth and breadth of experience." It might even be that the project manager is outsourced. The presence of contractors and other service providers, either part of the project team or not, presents important people-interaction challenges. Some suggestions follow.

- Remember, just as with the other key people interactions we have discussed, constant and nurtured communication is absolutely essential.
- Work early to address any cultural, technological, or other barriers that might impede this needed communication—make a plan to do this.
- Don't treat people not on the project team as second-class citizens. Value them. Care about them. Seek to understand their perspective. Your efforts will pay big dividends! Many of them could be project stakeholders.
- When dealing with contractors or people not directly on the project team, record-keeping and communication become even more important to avoid misunderstandings, re-work, and unwanted problems later on. Make sure important communications are in writing, and then follow up personally to discuss any questions or make clarifications. If changes are needed, put them in writing and get necessary approvals.

- Read all contracts thoroughly and make sure you understand them.
- If the other party is writing the contract, make a quick draft yourself of the contract before you see theirs. This will help you to have thought through what to look for and what questions to ask when you review their contract. After it is too late, many people wish they had more fully realized what they were signing!
- If you are writing the contract, ask the other party what they would like to see in it. Get input from other people who will be working on the project and people who have experience with contracts. Then, have a legal expert help you do the final contract to avoid problems you might miss. Make sure senior management signs-off on whatever the contract says. If you all work together, you are more likely to achieve win/win results.

People and Project Success

Notice how vital are the importance of people interactions and the perceptions of people, in addition to technical performance, as expressed in the following definition of project success from a study of approximately 650 projects (Baker, Murphy, and Fisher; 1974).

> "If the project meets the technical performance specifications and/or mission to be performed, and if there is a high level of satisfaction concerning the project outcome among key people in the parent organization, key people in the client organization, key people on the project team, and key users or clientele of the project effort, the project is considered an overall success."

This definition of project success places a large emphasis on perception—a people feeling. Notice also that schedule and budget are not explicitly stated as part of project success, although they undoubtedly somewhat affect perception and feeling. Nonetheless, such a definition certainly exemplifies the great importance of people in measuring and accomplishing successful projects.

Summary and conclusions

Good people interactions are vital to successful project management and must be carefully nurtured. These interactions permeate essentially every aspect of the project and are also the most time-consuming responsibility of the project manager. The above discussion should clearly demonstrate to the reader that people are indeed the most important key to successful project management. Because of the importance of people, a project manager needs to learn to sincerely care about each person associated with the project, and nurture these people interactions to achieve success. Charles Brod summarized it well, "People are project management!"

References

Baker, Bruce N., David C. Murphy, and Dalmar Fisher, "Factors Affecting Project Success," Ch. 35 of David I. Cleland and William R. King (eds.), Project Management Handbook, 2nd Ed., Van Nostrand Reinhold, New York, 1988.

Brod, Charles, Personal communication with the author, 1993.

Covey, Stephen R., The Seven Habits of Highly Effective People—Restoring the Character Ethic, Simon & Schuster, New York, 1989.

Flannes, Steven W. and Ginger Levin, People Skills for Project Managers, Management Concepts, Vienna, Virginia, 2001.

Gido, Jack and James P. Clements, Successful Project Management, South-Western College Publishing, Cincinnati, Ohio, 1999.

Larsen, Eldon R., "Not Another Meeting!—Planning and Conducting Effective Meetings," Proceedings of the ASEM National Meeting, October 13, 2001.

Larsen, Eldon R., "Start Projects Right—How to Develop Effective Project Objectives," Proceedings of the ASEM National Meeting, October 4, 2002.

Larsen, Eldon R., "Communicating With the Boss and Other Public Things," Management Topical: The Engineer in Transition to Management—Tutorial, Published in the Proceedings of the topical at the AIChE National Meeting, March 31-April 3, 2003.

Pinto, J. and O. Kharbanda, "Lessons for an Accidental Profession," Business Horizons, March-April 1995 (quoted by Gido and Clements, p. 83; 1999, cited in this paper).

Communication—A Better Understanding

Abstract ID: 481744
Paper # 330c
Dr. Eldon R. Larsen
Marshall University

Today's business organizations often seem less caring, more informal, and having less long-term mutual-commitment. Our interactions are ever-increasingly more digital. Hence, the need for excellent and clear communications between employees, managers, peers, customers, and the outside world will continually become more vital to long-term successful enterprises. This paper presents principles, concepts, and tools to help the reader gain a new vision of communication and how to do it better in today's challenging environments.

Additionally, this paper presents ideas on what is truly needed for good communication—what foundation concepts are necessary if we really hope to successfully communicate. The paper discusses how communication takes place and reviews the very important concept of *empathic listening* as defined by best-selling author Stephen Covey (2004). Finally, it is important to make a communications plan when we need to manage communication.

Introduction

Communication is one of the most important concepts and tools in human life. It is also one of the most used and abused words in human interactions. Its real meaning is rarely fully understood. Significantly, when one really understands the meaning of communication, a paradigm shift often occurs. With this paradigm shift, we start to look at our interactions with other people in a whole new, deeply meaningful way. We open the door to truly effective human interaction.

We use communication in essentially every interaction we have with other people, be it verbal, written, silent symbolism, body language, pictures, sounds, or other ways. Yet, it is also communication, miscommunication, poor communication, or lack of communication which frequently gets us into challenging situations and can cause frustrating emotions, at least when we communicate in the way that people commonly define communication.

In interviews with hundreds of project managers which my graduate students have conducted over more than a decade, many of these project managers state that some of their greatest challenges come from working and communicating with people. Why do we have so much trouble with communication? Why is it such a challenge? What are we missing?

Someone once said, "I know that you believe you understand what you think I said, but I am not sure you realize that what you heard is not what I meant." (Dinsmore, 1990) Does this sound familiar? If so, you are not alone. It happens all the time.

This paper presents important ideas and concepts that can help. One of the most important concepts discussed is what the word *communication* really means.

A better understanding of the root meaning of *communication* can cause a very important paradigm shift, to the point that the word takes on a whole new meaning. It can give us new vision. This help opens the door to better communication for us all.

Do We Really Need Good Communication?

Project managers spend the vast amount of their time communicating in some way or another. They are constantly providing direction, asking for information, clarifying objectives, seeking for understanding, and giving feedback.

Whether you are an employee or a new or seasoned manager, it is important for you to realize that (1) bosses want to be kept informed, (2) bosses don't like surprises, and (3) bosses depend on communication to help them to plan and to organize, motivate, and control or influence people—actually the same could be said for most of us, not just bosses. Good communication is like the lifeblood of a successful project manager or a successful leader.

Project managers, team members, stakeholders, and others all struggle when we use poor communication techniques. Communication in its many-varied facets consistently comes up as a major managerial issue.

It is commonly stated that the three most important factors to consider in real estate are (1) location, (2) location, and (3) location. In management, especially in project management, the three most important factors to consider are (1) communication, (2) communication, and (3) communication.

What Is Communication?

What does the word *communication* really mean? Why ask this question about a word so commonly used? Everyone knows what we mean by communication—don't they? *Do we*? Let's look a little deeper.

High school English teachers often tell their students to look at the "roots" of a word in order to really understand it. These teachers often split words into various prefixes, suffixes, and main-word fragments, and then look at the Latin or other linguistic meanings of these word fragments to determine the true meaning of words.

Let's do this root-understanding-exercise for the word *communication*. We will look at the roots in French, rather than Latin, because the roots form common French words. *Communication* is in fact a French word derived from Latin, and is spelled the same in both French and English.

Comm/un/ication

First, we break communication into three word fragments: *comm, un,* and *ication*. *Comm* comes from *comme*, which is a separate French word that means "Like" or "As." *Un* in French means "One." Finally, *ication* essentially means "The process of becoming." When we put this all together the fundamental meaning of the word comes to light:

Communication = The process of becoming "like one" or "as one."

Let's say it again: Communication means the process of becoming like one or as one! It is the process of becoming unified!

Do we realize what this is saying? Think about the implications of the root-word definition! Communication is not just sharing or transmitting information. Communication is not just sending out a message. Communication is not just giving instructions. It is not just telling someone what you think or know. In fact, in and of themselves none of these are true communication according to the root-word definition. They can't be; they are one-way.

By the root-word definition above, communication almost always must be at least two-way. How else can we become like one? To be effective, this two-way interaction often needs to be very interactive and iterative. The only exception is the rare case when the receiver of the information just happens to fully agree with the sender of the information. Even then how do we know that communication has been achieved, unless we both interact and *communicate*?

Note that "The process of becoming like one" does NOT mean becoming the "same." We can be unified without being the same. In fact, we will be stronger and can synergize more effectively if we value differences, and yet are unified. It is a bit like having two pieces of a two-piece puzzle. The two pieces are not the same, but when they fit together and become *like one*, then they make a clear and complete, unified whole picture. Separately they are greatly lacking. We need to seek to become unified for project success.

In summary, true communication is a process of coming-together, in common understanding, vision, feeling and purpose, so that we truly understand each other to the point that we are becoming like one.

Principle:

Communication = The process of becoming like one

For most people, this concept, this original root-word definition, or could we even call it a *principle*, is a paradigm shift. It is a whole new way to look at what we are really trying to do when we say we want to communicate. It changes the whole picture or vision of what we are trying to accomplish. It changes our whole approach. The objective is not to just shove our opinions down someone else's throat! *We are trying to become like one.*

As a side note: If you physically try to shove something down someone's throat, you will likely get a response like gagging or vomiting. Think about that next time you try shoving your opinion down someone else's throat—it simply won't be pretty! It won't work, no matter how well intentioned you might be. This is not the way to communicate.

With this new vision—that we are seeking to become unified—we are now ready to realize that our very motivation in communication must change. Communication is not a one sided paradigm; it is two-sided at least, and it is serious and amazing. It involves us in a whole new perspective, not just our own perspective, and for a whole new purpose. How will we start communication anew with this new paradigm? We now are ready to investigate this more.

What Is Needed to Truly Communicate?

On what foundation are you to build if you want to truly communicate, if you want to become like one or as one with someone else? There are two main foundation building blocks I will discuss here.

1. Honestly and Mutually Care About Each Other

First, to truly communicate, you, and those with whom you are attempting to communicate, need to honestly and mutually care about each other. You must mutually respect each other and be considerate of each other's needs and perspectives. Without this mutual caring, communication efforts will quickly deteriorate to our unfortunately more modern one-way transfer of information we incorrectly call communication.

This presents a new challenge: Since you can't control the other person with whom you are trying to communicate, can this idealistic *communication* really happen? Isn't this a lost cause?

The answer is that you must start with yourself—only you can decide that you will really care about the other people around you, not just for your own self-interest, but for theirs too.

When you start with yourself, when you change your own motivations, almost miraculous interactions start to happen. For some strange reason, others seem to start to care about you too! They want more to work with you and to desire the oneness in purpose and vision and understanding that you emanate and that you too desire.

Thus, communication becomes a very personal issue. It is not an issue of better technology. "Communication among people depends not on technology, but rather on forces in people and their surroundings. It is a process that occurs within people." (Gibson, et al., 1997)

2. Mutually Align

Second, to truly communicate you also need to mutually align. Alignment means that you and those with whom you work all get pointed in the same direction, so to speak. You work toward the same goals, see the same vision and have a desire to work together toward achieving that vision. See Figure 1. One big question is "How do we accomplish this alignment of people?"

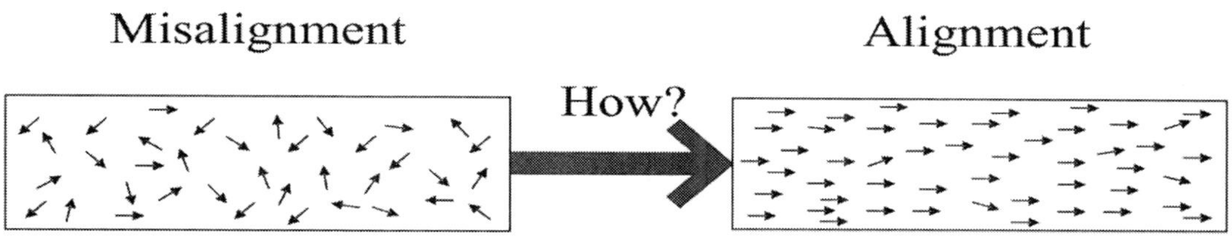

FIGURE 1 Transforming People from Misalignment to High Alignment

Let's look at an analogy to better understand personal alignment needed for better communication. When we take an automobile in for an alignment, Figure 2, what we mean is that we want to get all the wheels pointing in the same direction within certain tolerances. If the wheels are in poor alignment, the consequences are increased tire wear, steering difficulty, lower fuel efficiency, and extra wear on other parts of the car. Good wheel alignment is very important!

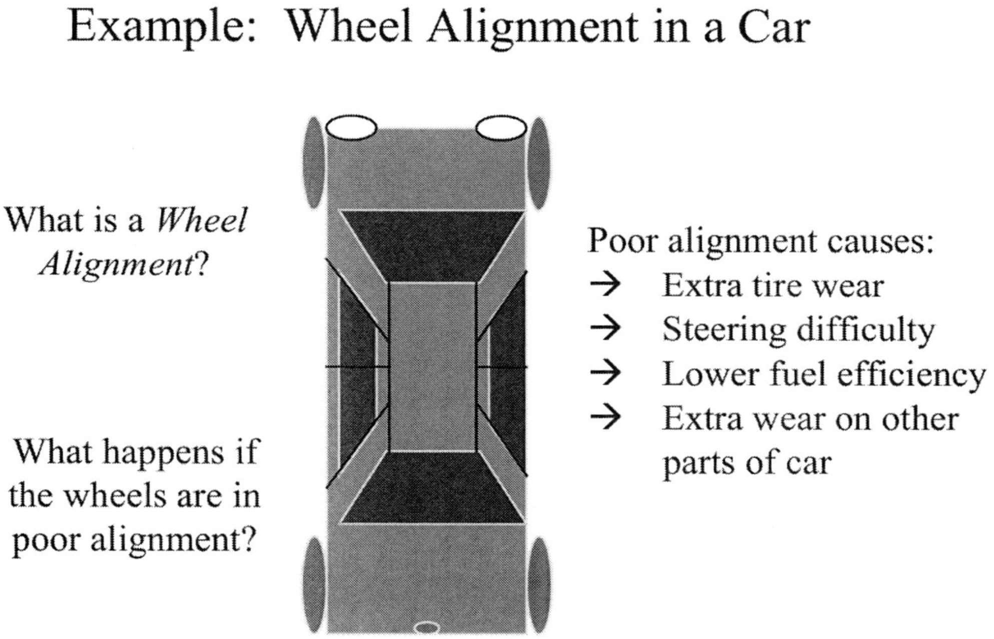

FIGURE 2 Good Alignment is Important for Cars—It is for People, Too!

Alignment of team members toward project objectives is equally important. Like the wheel-alignment analogy, poor team member alignment can lead to extra wear on the team members, difficulty to steer or manage the team, poor work efficiency, and even eventual wear on the larger organization. In contrast, a well-aligned team runs well, is much easier to manage, effectively and efficiently carries out the team's work, and improves the well-being of the organization as a whole.

Another analogy gives some insight into various ways we might try to achieve alignment. In this case, Figure 3, we use an iron bar that is not magnetized. In this bar the iron atoms are all spinning with dipole moments pointing randomly in an infinite number of directions. If we were to align enough of these iron atoms' dipoles, by use of some form of work, we could magnetize the iron bar.

This magnetization could be done by electrical induction, where we coil an insulated electric wire around the bar and pass a direct current through the wire. It could also be done by placing a permanent magnet next to the iron bar, with the work to align the previously random iron atoms coming from the permanent magnet. A third way to magnetize the iron bar would be to position the bar in a north-south orientation and pound on it with a hammer.

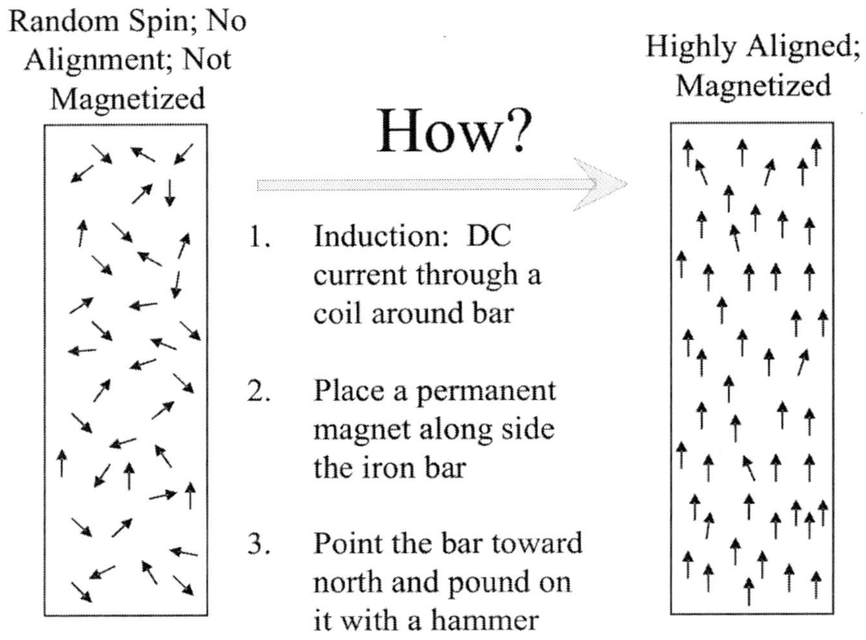

FIGURE 3 Alignment of Atoms in an Iron Bar

A group or team of people might be analogously aligned like the atoms in the iron bar. The team analogy to the iron bar is shown in Figure 4. Again, there are several ways that we might try to enhance team member alignment.

Similar to induction, we could put the team under the influence of some good interpersonal training and team-building exercises. Under this influence, the team is *induced* to start making better efforts to work together and to better understand each other.

Alternatively, we might expose the team to the example of people who have learned to permanently align themselves in order to achieve group objectives. These example-emanating people could show our team how to align by their good examples, or they might even become members of our team or work alongside us as mentors or partners. By their example and leadership, we learn how better to communicate and unify.

A less desirable approach would be to have the boss come in, point us in the right direction, and then "hammer" us into alignment by, for example, threatening us with dire consequences if we don't align.

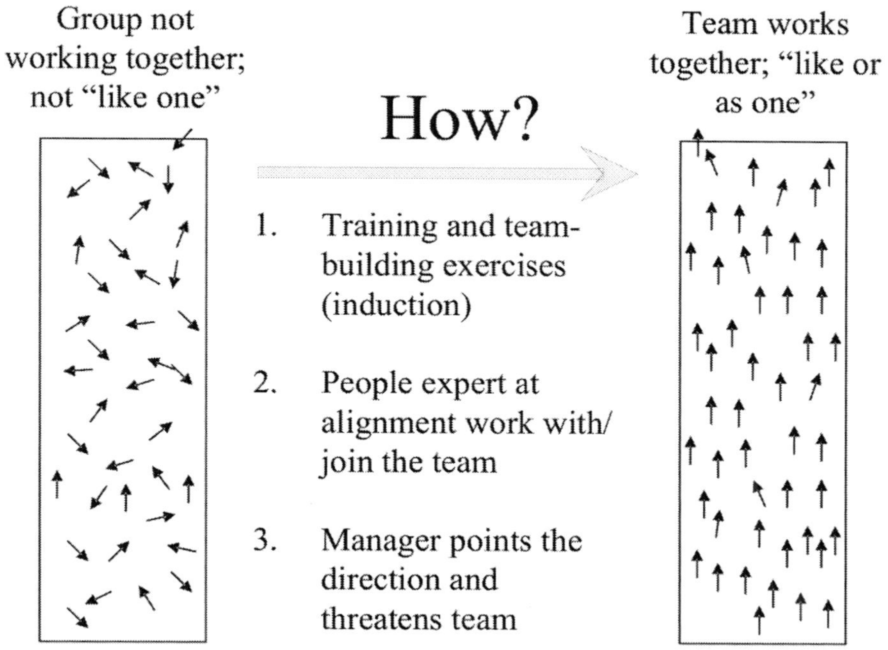

FIGURE 4 A Group of People Aligns into a *Team*

Must a group or team be perfectly aligned? No. In fact, a small degree of misalignment can have a few important benefits: (1) different perspectives, and the raising of concerns and alternatives, often aid our understanding, (2) creativity can be enhanced as we learn to accept and value differing perspectives, (3) strength can come from learning to manage minor personal-interaction challenges, and (4) people are more likely to feel they are in an environment of acceptance and freedom to communicate openly when their differences of opinion are valued.

There is a fifth benefit to having some lack of alignment: you might find, as you try to communicate with those who are not aligned, that there are some very important reasons those people are not aligned. For example, they might know something you don't, but they have been hesitant to share their knowledge. Through your sincere, caring communication efforts, you might very well find that the majority of the team is not pointed in the correct direction and that the few misaligned people actually were headed where the whole team really needs to be going!

On the other hand,, too much misalignment can cause (1) increased dissention, (2) decreased team efficiency, (3) decreased team effectiveness, (4) more arguments and backbiting, (5) lack of team spirit, and (6) increased chance of destructive behavior, and perhaps even attempts to sabotage the project. All of these problems are signs we are not really communicating—we are not becoming "like one." Therefore, high levels of misalignment in teams or individuals within a team must be dealt with, hopefully in a way that achieves better communication. Good use of the principles and concepts in this paper will help. If, after sincere application of these concepts, the matter cannot be rectified, *then* we might need to look at more drastic actions of removing a person from the team; but this should be a last resort, as this can have many negative consequences.

How Does Communication Take Place?

There are various modes and mediums through which we communicate. Examples are written words, speaking, body language, face-to-face meetings, telephone, voicemail, electronic mail, letters, memos, video-conferencing, and groupware. Of the above, *speaking* is one of the most common and important. Speaking is also a mode of communication where we often have difficulty. For example (Dinsmore, 1990), do you understand what former presidential aid Ron Ziegler meant in this statement:

> *You should not interpret by my use of "least likely" that ultimately, or when the final decision is made, that that may not be the decision, but what I am saying is that it is only one of the matters under consideration and the decision has not been made.* (Rees, 1979)

The above quotation made sense to the speaker, but is a little difficult for most readers or listeners to understand. How can we do better? Is there some model that will help us to understand better the process by which communication takes place? If so, it might give us an idea of how to improve communication.

The Effective Communication Model

One communication model that is found in many textbooks is called the Communication Process. (Gibson, et al., 1997) In this model there is a Sender who says some message in a certain way (medium) to a Receiver. The Receiver in turn gives feedback, indicating what effect the message had on the Receiver. The sending and feedback paths form a loop. This is shown in Figure 5.

Between various steps there are events involving encoding and decoding of the messages, which can muddle the message somewhat. These encoding and decoding events can be filtered or selectively influenced by personal perspective, bias, emotion, lack of understanding, errors, etc. Along the path there is also background noise caused by such things as disturbances, interruptions, side conversations, static, etc.

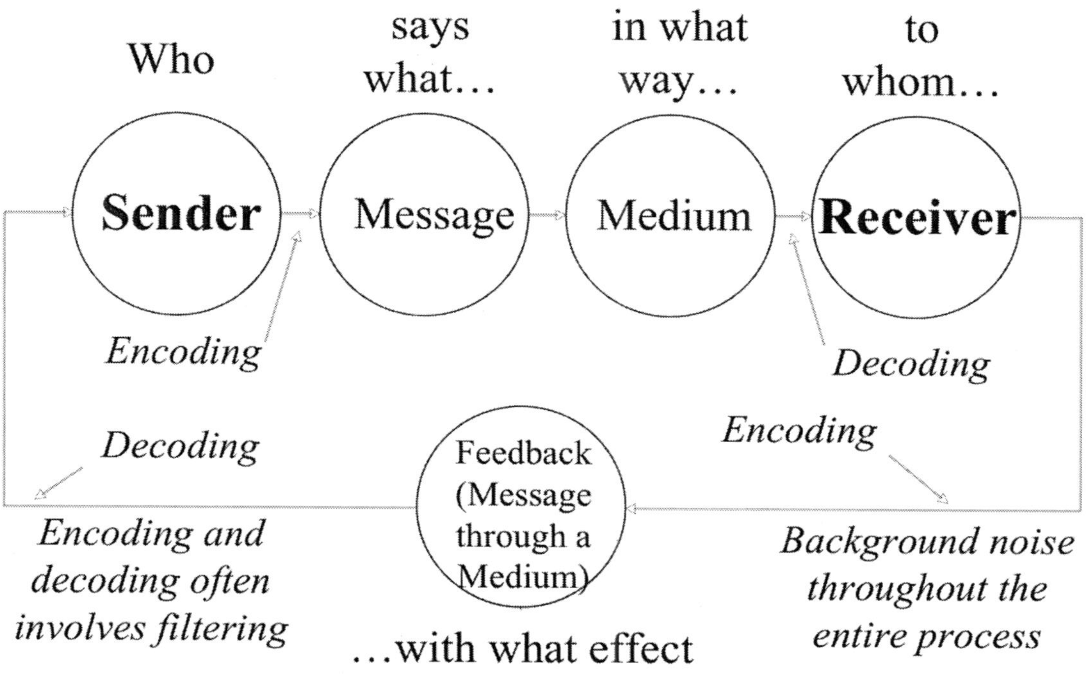

FIGURE 5 A Very Common Communication Process Model

At first glance the Communication Process model seems adequate, but in reality it tells only half the story. Communication is not just one way with some feedback—this is not how we usually become "like one" or "as one."

Instead, a more correct communication model shows that both of the people in the communication loop have information to share, both need feedback, and they must interact until they come to a common understanding—until they become like one. Figure 6 shows this improvement in the model, which I call the *Effective Communication Process Model*, which includes two-directional interaction. It is founded on principles of *mutual caring and alignment.*

To summarize, communication takes place through two-way interaction, with all participants having the opportunity to give their views as Senders and to give feedback as Receivers. In addition, effective communication must be founded upon mutual caring and alignment.

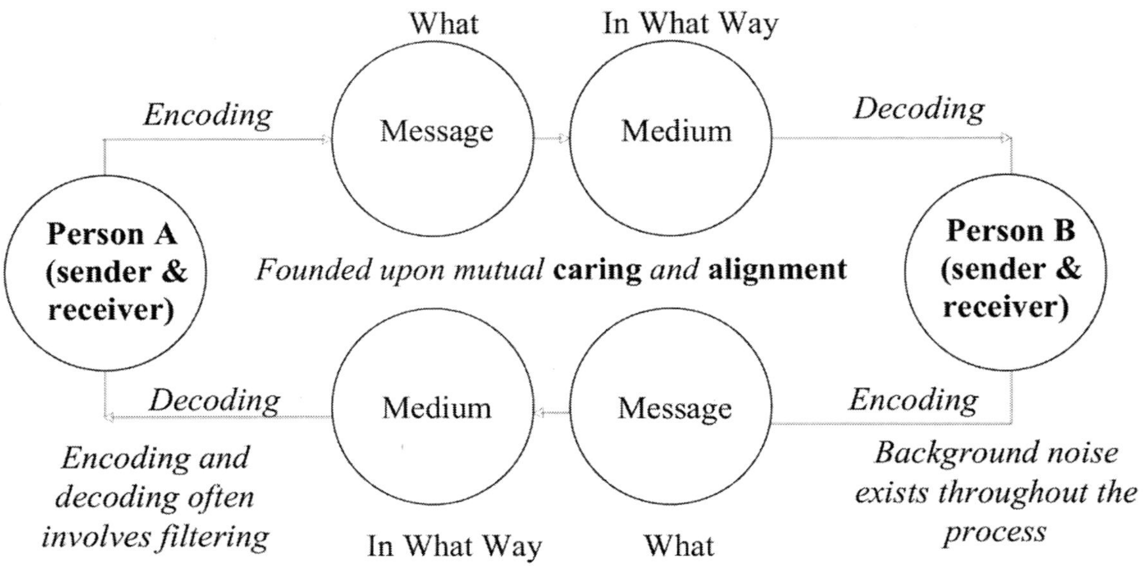

FIGURE 6 Two-Directional Effective Communication Process Model

Empathic Listening

Jack Gido and James P. Clements truly say, "The heart of communication is not words, but understanding. Not only to be understood, but also to understand. Half of making communication effective is listening." (Gido and Clements, 1999)

Much has been written about how to listen better, particularly on the topic of so-called Active Listening. But one of the most important listening concepts, when it comes to true communication as defined above, is the concept of *Empathic Listening*. (Covey, 2004)

What is Empathic Listening? Empathic Listening is not Active Listening, which is skill based. Instead, Empathic Listening is character and principle based. It is Active Listening along with sincerely caring about the other people with whom we are trying to communicate.

Empathic Listening = Active Listening + Sincere Caring

In Stephen Covey's bestselling book *The Seven Habits of Highly Effective People*, the fifth of the seven habits is

Seek first to understand, then to be understood." (Covey, 2004)

This habit is key to successful communication! We need to care so much about the other people with whom we are trying to communicate that we are willing to try to understand them, even before we try to have them understand us. We do this to increase *our* understanding, to get a different perspective, to learn new information, and to show the other person we truly care about him or her.

In addition, by understanding the other person, we greatly increase the possibility that we can synergize with them—to take the best of both of us and achieve things neither of us could have done by ourselves.

For many people this is a major behavioral change, but it's worth it! This change brings rewards far beyond most people's expectations—almost miraculous changes and rewards in some cases. Empathic Listening is probably the most important part of Covey's fifth habit—"Seek first to understand, then to be understood." Again, this form of listening must be built upon sincere caring for other people.

Covey has said, "If you use techniques or skills alone, I may sense duplicity, manipulation, or insincerity—I might wonder about your motives... I might feel over time that you really don't care so much about me as what you can get out of me."

Empathic Listening is based on truly caring about others. Someone once said, "If you really want to understand me, don't just listen to what I say, listen to what I feel."—This adds real meaning. Understanding other people, and then seeking to have them understand us, is vital to achieving the goal of becoming like one or as one.

To summarize, we need to Empathically Listen and seek first to understand, then to be understood.

Poor Listeners versus Good Listeners, and Miscellaneous Communication Barriers

Paul Dinsmore lists some typical characteristics of poor listeners and good listeners as follows: (Dinsmore, 1990). These are summarized in Table 3.1. Notice that you, the listener, are in control of most of the above good or bad listening habits!

Watch out, there are even more barriers to effective communication. For example, when trying to communicate, consider the following barriers in addition to those already listed above (some of the following are taking from Gibson, et al., 1997):

- Value judgments
- Semantics problems
- Jargon—technical, business, etc.
- Low credibility—low trust & confidence levels
- Proxemic behavior—physical distance between individuals (0-1.5 ft = intimate; 1.5-4 ft = personal; 4-12 = social; >12 feet = public)
- Assumptions
- Time pressures
- Differences in perception
- Emotions, either preceding or during communication
- Selective listening—daydreaming or preoccupations
- Thinking of counter-arguments while someone else is speaking
- The communication *medium*—telephone, email, etc.

- English as a second language
- Memory
- Distractions
- Too much formality
- Too big of group to properly interact to achieve unity

In spite of the fact that most of the above barriers are somewhat familiar to most of us, we still struggle with communications. Klaus Krippendorf says, "Despite the tremendous advances in communication and information technology, communication among people in organizations leaves much to be desired." (Gibson, et al., 1997)

Table 3.1: Poor versus Good Listeners (Dinsmore, 1999)

The Poor Listener:

- Always interrupts
- Jumps to conclusions
- Finishes my sentences
- Is inattentive; has wandering eyes and poor posture
- Changes the subject
- Writes everything down
- Doesn't give any response
- Is impatient
- Loses temper
- Fidgets with pen, pencil, or paper clip nervously

The Good Listener

- Looks at me while I'm speaking
- Questions me to clarify what I'm saying
- Shows concern by asking questions about my feelings
- Repeats some of the things I say
- Doesn't rush me
- Is poised and emotionally controlled
- Reacts responsibly with a nod of the head, a smile, or a frown
- Pays close attention
- Doesn't interrupt me
- Keeps on the subject until I've finished my thoughts

Plan to Communicate

In order to put effective communication into action, one additional tool can be very helpful: a communication plan. Someone once said, "There are three kinds of people in the world: those that make things happen, those that watch things happen, and those that wonder what happened." A plan helps you to "make things happen."

When making a communication plan, include (1) what to communicate, (2) with whom to communicate, (3) frequency and when to communicate, (4) format (memos or reports, for example), (5) how to communicate—the medium: formal or informal meeting, email, newsletter, voicemail, etc., and (6) who will do the particular communication effort. When making the plan, seek input from the key people who will likely be involved in the needed communication process. Their input can prove invaluable by helping you to understand their needs. A simple format for developing a communications plan is shown in Figure 7.

COMMUNICATIONS PLAN FOR PROJECT:______________________					
What to Communicate?	To Whom?	Frequency?	Format?	How? (Medium)	Who Will Do It?

FIGURE 7 Example Format for a Project Communications Plan

Summary

In summary, remember: (1) good communication is critical to success, (2) communication is the process of becoming "like one" or "as one," (3) to communicate well, you have to truly care about other people enough to sincerely want to understand them, (4) "Seek first to understand, then to be understood"—empathically listen, and (5) make a communications plan and use it!

References

Covey, Stephen R., *The Seven Habits of Highly Effective People—Restoring the Character Ethic*, Simon & Schuster, New York, 2004.

Babcock, Daniel L. and Lucy C. Morse, *Managing Engineering and Technology—An Introduction to Management for Engineers*, 3rd ed., Prentice Hall, New Jersey, 2002.

Dinsmore, Paul C., *Human Factors in Project Management*, revised edition, AMACOM, American Management Association, New York, 1990.

Gibson, James L., John M. Ivancevich, and James H. Donnelly, Jr., *Organizations—Behavior, Structure, Processes*, Irwin/McGraw-Hill, Boston, Massachusetts, 1997.

Gido, Jack, and James P. Clements, *Successful Project Management*, South-Western College Publishing, Cincinnati, Ohio, 1999.

Mintzberg, Henry, *The Nature of Managerial Work*, Harper & Row, Publishers, Inc., New York, 1973, p. 39.

Rees, Nigel, ed., *Quote…Unquote*, St. Martin's Press, New York, 1979, p. 79.

Shainis, Murray J., Anton K. Dekom, and Charles R. McVinney, *Engineering Management—People and Projects*, Battelle Press, Columbus, Ohio, 1995, p. 33.

Planning and Conducting Effective Meetings

Abstract ID: 481745
Paper # 330d

Dr. Eldon R. Larsen
Marshall University

Meetings can be viewed as necessary evils, valuable experiences, inspirational events, or intolerably boring curses. The very word *meeting* can cause various emotions to arise within us. This paper presents various rules and suggestions that lead to more effective and more enjoyable meetings. The approaches are founded on common sense and solid principles of management and human relations that are frequently overlooked. The rules and suggestions focus on reaching meeting objectives, effectively using agendas, nurturing good human relations, and ending meetings with an effective plan of action.

Introduction

Many people abhor meetings. Perhaps you have had similar feelings; many people do. Why do people dislike meetings? Common complaints are that meetings are too long, not well organized, pointless, and a waste of time. Some of the major complaints (Dressler, 1996) about meetings are as follows:

- Their purpose is unclear
- Participants are unprepared
- Key people are absent or missing
- The conversation veers off track
- Participants don't discuss issues, but instead dominate, argue, or take no part at all
- Decisions made during the meeting are not followed up

A survey of 78 project management professionals indicated that meetings were the major time consumer in their daily schedule, those meetings taking up an average of slightly less than 2 hours per day. "The survey showed that the participants attended over six meetings per week. In those meetings an average of 25 percent of the time was spent engaging in off-the-track banter, generalized complaining, lobbying for pet causes, and other unproductive activities. When respondents were asked, 'What do you think contributed to the eventual non-productiveness of the last meeting you attended?' the three most prevalent responses were (Dinsmore, 1990, 139-140)

- Improperly planned meeting (no written objective or detailed agenda)
- Inept leadership
- Undisciplined participants"

When experience with meetings is not good, people can actually get a negative conditioned response to the very word *meeting*. Yet, meetings can and do play an important part toward facilitating progress and promoting understanding among people—a key to successful management and leadership. How can this be accomplished more effectively?

Rules and Suggestions

To plan for and to conduct effective meetings, project managers and meeting participants generally need to follow ten basic rules. When used *appropriately*, these rules lead to success. They are founded upon solid principles of management and good human relations.

Rule One. *Clearly identify the meeting objective. And, if the meeting is formal, write the objective and distribute it to all concerned in advance.* This is a vital, and perhaps the most important, rule for conducting effective meetings.

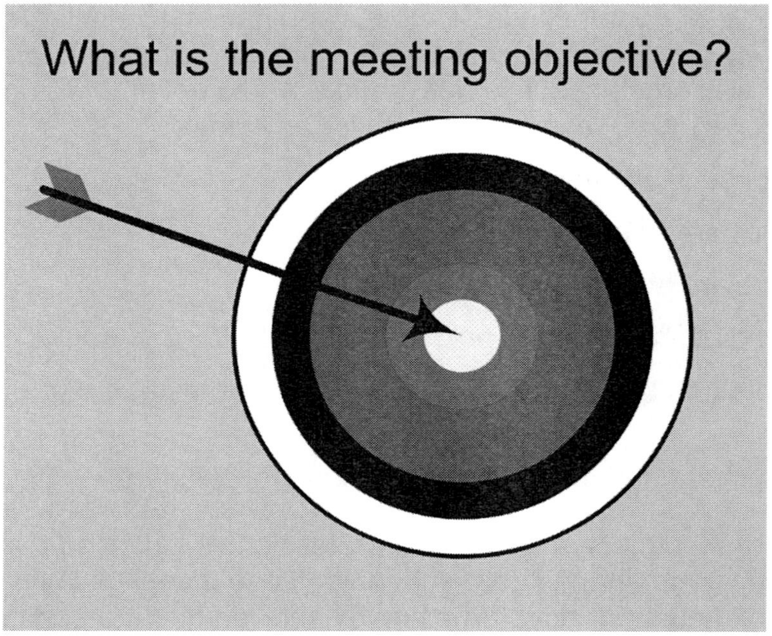

FIGURE 1. Clearly Identify the Objective of Every Meeting

Each semester in my graduate project management courses, I ask my students if they are given the objectives of the meetings they are asked to attend. For every one hundred students, I normally have less than five who are given the meeting objectives. This is disturbing when one considers that the majority of my students are full-time employees in industry, where productivity is of great concern.

When deciding to hold a meeting, a leader must determine among other things why the meeting is important and what is to be accomplished. The leader should ask, "Is this meeting necessary?" and "What is the objective or purpose of the meeting?" Answering these questions will essentially answer whether it is also necessary to meet. Though the meeting objectives might seem obvious to the leader, they frequently are not obvious to the participants. If the meeting objective is obvious, then it should not be difficult to write that objective; yet this could indeed be a challenge to many meeting participants. If you can't write down the objective, you don't truly understand it.

Principle:

If a meeting doesn't have a clearly understood purpose, don't have it! If it is going to waste people's time, either change it or cancel it!

Some generic areas around which meeting objectives might be formulated include, for example,

- Solving problems
- Brainstorming new ideas or solutions, etc.
- Making decisions
- Sharing information (this can be a weak reason to hold a full-group meeting, since much information sharing can be done through e-mail, personal conversations, or sub-group meetings)
- Giving feedback (again, this too can be a weak reason to hold a full-group meeting, as much feedback if likely to more appropriate on a one-on-one basis)
- Planning
- Selling
- Teaching
- Conducting design reviews

Before we leave the first rule for effective meetings, there is an important point to understand. Objectives are much more effective if all key people involved agree generally on those objectives. Taking time in advance of the meeting to clarify and reach agreement on meeting objectives with all key players can go a long way toward avoiding difficulties in the meeting itself. A few advantages of doing this include

- Everyone is focused on what is to be accomplished

- Participants are less likely to come with hidden agendas
- The meeting agenda will more likely be clearly oriented to accomplishing the meeting objectives
- Agenda items not contributing to the meeting objective can and should be dropped
- The agenda items can be more easily prioritized
- The participants have a greater sense of purpose and understanding—they will behave more like a team

Rule Two. *Clearly define the meeting agenda, focused solely on accomplishing the meeting objective.* For formal meetings this agenda normally should be agreed upon in advance by the meeting participants, along with the meeting objective.

The meeting agenda should not include items that do not contribute to the meeting objective. Usually items that do not contribute to the meeting objectives can easily be communicated by some other means, such as one-on-one interaction, e-mail, or telephone conversations. Many such items also do not involve all the meeting participants anyway. Invite only those who will benefit directly from, or contribute directly to, the meeting.

Effective agendas include not only the topics to be discussed, but also who is expected to lead each topic's discussion or presentation and how much time is allotted for that agenda item. Allocate and prioritize time based on importance first, and then urgency—if an agenda item is not important, drop it. Urgency does not equal importance. As Covey (2004, 150-151) explains, "*Urgent* means it requires immediate attention…Urgent things act on us…We react to urgent matters…*Importance*…has to do with results. If something is important, it contributes to your mission, your values, [and] your high priority goals." For an excellent discussion on the difference between *urgent* versus *important*, see Covey (2004, 149-168).

Use wording to describe clearly the agenda items so that participant understanding is well focused. To further clarify an agenda topic, the leader might want to consider writing down the desired expectations next to each agenda topic. For example, if an agenda item is "Forecasts of business income," the leader might add the expectation note that "Please be prepared to decide as a result of these forecasts whether to increase production of XYZ." Making expectations clear will greatly enhance the effectiveness and efficiency of the meeting.

The items on the agenda should be prioritized with the most important items being covered as early as possible. Then, if the meeting appears to be running behind, the less important agenda items can be postponed to a later meeting without serious consequences.

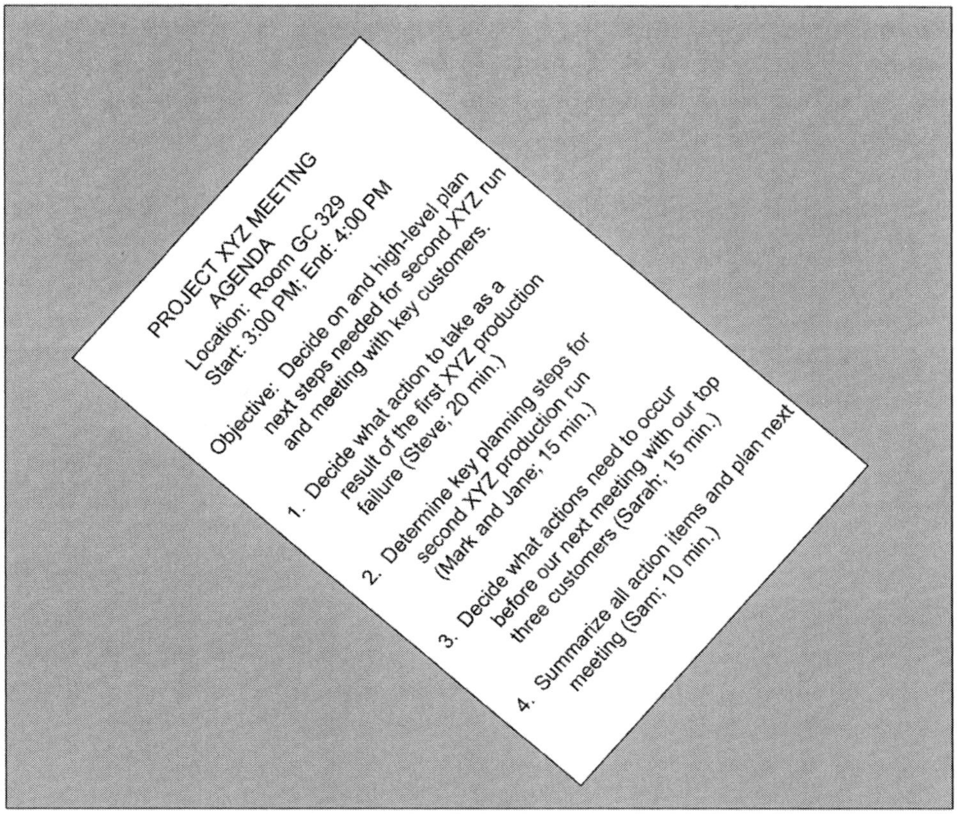

FIGURE 2. Clearly Define the Meeting Agenda to Accomplish the Objective

Be sure to schedule enough rest breaks as part of the agenda. These breaks should occur at least every one and a half hours or less, as longer periods between breaks will tend to decrease meeting effectiveness. Ten-minute breaks are usually sufficient, and should be no more than fifteen minutes except in unusual circumstances. Start promptly at the end of the break!

When distributing the agenda, include any needed background or supplementary reading material and data that will help participants prepare effectively for the meeting. Do not include material that does not directly contribute to the meeting objective.

Rule Three. *Have clear start and end times for the meeting, and treat these as a contract.* Most meetings have a clear start time. It should be firm—start on time! Doing otherwise will lead to poor meeting behavior and punish those who try to be on time.

Unfortunately, many leaders do not plan meetings to have a clear end time. This is a mistake made by meeting leaders who are not sensitive to the courtesies that ought to be shown to other participants in the meeting—their time is important and they, too, need to plan their day. To do otherwise will lead to frustration for those involved.

Once the meeting agenda has been determined and it includes time allotments for each agenda item, there should be little difficulty in deciding on when the meeting should end. Wise meeting planners include a little "contingency time" toward the end of the meeting; but this is *not* for sloppiness—it is for unforeseen events which honestly could not have been planned for. Of course other factors must also be considered such as travel schedules, meal times, and so forth. But, once the schedule is set, stick to the overall schedule. This means don't overrun the end time—you have a contract, treat it that way. On the other hand, few people, if any, will mind if the meeting ends early.

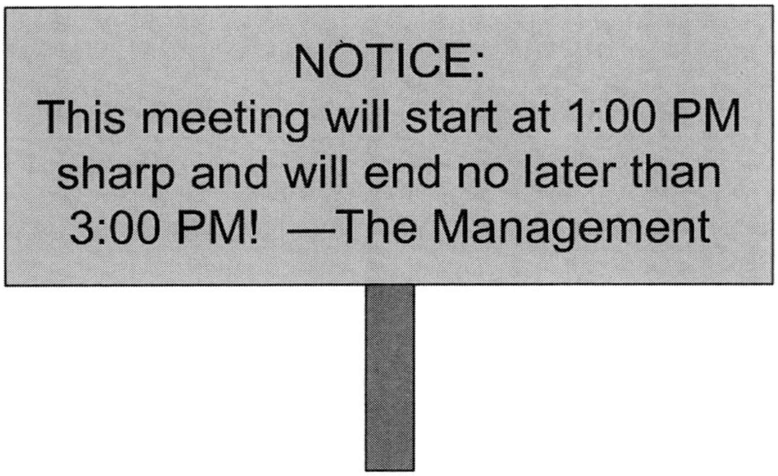

FIGURE 3. Have Clear Start and End Times, and Treat as a Contract

This does not mean that individual agenda items cannot be adjusted mid-stream. The agenda is simply a plan to reach the meeting objectives. The plan is your servant, not your master—it is there to help you! Some adaptation of the agenda might be needed to reach the objectives. Caution is in order, however—the agenda should not be so flexible that control and discipline are lost and the meeting objectives are not reached.

A final note on the time contract: It is common for the meeting to approach the end time with agenda items still to be covered. As pointed out already, the consequence of dropping off the last few agenda items will be small if the agenda has been prioritized. Still, there is frequently a temptation, especially among meeting leaders, to feel that they must get through the agenda and, therefore, they don't want to quit. The decision to run over time is *rarely* a good one. Some consequences of running overtime include

- Loosing participant attention
- Hastiness leading to poor decisions
- Loss of trust for future time contracts
- Ill will toward the leader

Sometimes we feel that most of the participants will not mind staying beyond the stop time, and we can let the one or two people who need to leave do so. While this may indeed be the case, remember that there is a reason for all participants to have been invited, including those one or two who can't stay. Otherwise they shouldn't be there. When you tell someone that it is all right for them to leave and the rest of the group will continue, you basically say, "We don't need you," or else, "We no longer need your input." Even if you state that you don't want them to get these messages, they frequently will.

This can cause feelings of loss of respect and can cause resentment, even though these results are not intended.

People who are team players will want to be involved with the discussions and decisions—don't leave them out. Conversely, the remaining team members might well feel resentment at the pressure to stay in the meeting when others are allowed to leave. It is nearly always better to keep the time contract and to schedule another follow-up meeting if it is truly needed. There may be rare exceptions, but the key word here is *rare*.

Rule Four. *Be prepared for the meeting, both in content and in process.* It is the responsibility of the meeting leader to assure that the meeting will be valuable for the participants. Proper pre-meeting planning and preparation for both content and meeting process are essential steps leading to success in making the meeting valuable.

First, *content preparation*. The agenda should be issued enough in advance so that each participant will have adequate time to be content-prepared as needed for his part. Participants should be expected to come prepared with any useful visual aids, handouts, demonstrations, reports, data analysis, and so forth, as appropriate. Use of visual aids should be encouraged, since they will help participants to better communicate and to stay focused on their assigned topic without rambling. In addition, visual aids usually greatly help in understanding.

Meeting participants should expect each other's assignments to be completed on time, and they should be prepared themselves. The expectation of disciplined preparation should be taught and used well. However, the leader should remember that this is a two-way responsibility and that he must do his part in working with the other participants in advance in a win/win approach so that these expectations can be achieved—help them prepare in any way that is needed. The leader especially must be prepared or the others will quickly see hypocrisy and often choose to follow the leader's bad example.

Additional step to take before the meeting, which will improve content preparation, include

- Consider the need for some participants to make *pre-presentations* to key decision makers before the meeting; some presentations might need to be made before the meeting to key people to prevent embarrassing surprises, to enhance time for understanding and communication before the meeting begins, and to get early feedback and buy-in. Sometimes this is called *pre-positioning*. It can be quite powerful in making meetings more effective. Meeting energy will be more focused, decision-making will be more efficient, and deadlocks will be fewer or overcome.

- As needed, determine and clarify the roles and responsibilities of all participants. This could be done one-on-one between the leader and each participant before the meeting. Some example roles and responsibilities are given in Table 4.1.

- Engender expectations among meeting participants that the meeting success will be measured mainly by how well the meeting objectives are achieved in a matter satisfying to the participants.

- Anticipate potential interpersonal communication challenges and work toward alleviating these *before* the meeting.

- Assure that all handout material is prepared in an organized way for all participants.

Second, *process preparation.* Process preparation involves assuring adequate facilities and supplies are provided to make a comfortable meeting, as well as being prepared to run the meeting in a pleasing way. There are many considerations with respect to being process-prepared for the meeting.

You should plan the meeting in a way that will maximize participation toward reaching the meeting objective; when possible, consider trying to decrease presentation time and increase time where all can participate together.

The following process items, as appropriate, should be checked well enough in advance of the meeting so that changes can still be made as needed.

- Make meeting room arrangements, including, but not limited to:
  - ✓ Reserve the right-sized meeting room and additional rooms as needed
  - ✓ Assure room temperature is properly adjusted and will not be distracting; find out *in advance* how to get it adjusted if needed
  - ✓ Assure seating arrangement fosters participation; can squeaky seats be oiled?
  - ✓ Test visual-aid equipment in advance to assure it works properly; focus projectors; check visibility from all participant locations
  - ✓ Assure miscellaneous supplies are available, such as tape, markers for white boards and flip charts, sticky notes, pencils, notepads, extension cords, flip charts, video-tape player, lap-top computer, push-pins, name-tags, microphones, projectors and screens, erasers, and pointers
  - ✓ Arrange for refreshments as appropriate, well in advance and then confirm
  - ✓ Assure adequate trash containers are available
  - ✓ Check for adequate lighting and learn to adjust it as needed
- In advance of the meeting, not during, arrange for someone to take minutes and clarify expectations for those minutes
- Are there other environmental distractions that could cause participants difficulty in paying attention or make the meeting seem too formal and not relaxed? Noise from adjoining facilities or from in-room heating and air-conditioning equipment, for example, can be unexpectedly detrimental.
- Understand the decision-making authority of the participants; find out who has authority relative the meeting objectives and find out how the group can tie into that authority quickly and effectively if need be.
- Understand what financial arrangements need to be made for the meeting and how these will be executed. Then, make sure the arrangements are made.

Table 1: Some Example Roles and Responsibilities
Example *Group-Member* Roles: • Be prepared; fulfill all assignments on time • Empathically listen; Communicate, communicate, communicate • Help plan group work • Set objectives for and plan work in alignment with the group objectives and plans • Help establish and then obey group conduct rules • Actively contribute toward good teamwork • Be clear, concise, and don't dominate; but be a valuable contributor
Example Leader Roles (in addition to Group Member): • Lead, but don't dominate • Be responsible for meeting objectives and agenda • Complete all pre-meeting arrangements • Build teamwork environment; praise contributions • Work to empower group members • Set the example in true communication • Lead the team in reaching the meeting objectives • Be responsible for meeting follow-up, help people account for responsibilities
Example Minute-Recorder Roles (in addition to Group Member): • Summarize the essence of important discussions, decisions, and reasoning • Avoid direct quotations with names • Clearly summarize discussion perspectives, to avoid re-hashing of these points • Issue minutes promptly to all invitees, following approval by the leader
Example Facilitator Roles, sometimes the Leader, or another group member: • Help plan the meeting to maximize participation and reaching meeting objectives • Seek to decrease presentation time and increase team participation time • Arrive early; keep the discussions on track; eliminate side conversations • Empathically listen and help others to do the same • Value the differences in people; then protect people from being attacked • Help establish and enforce meeting rules agreed-upon by the group • Facilitate processes for taking needed actions • Assure clarity of decisions and assignments, who will do them, when they are to be completed, and to whom to report their completion • Help leader maintain proper meeting environment and assure all participate • Be the timekeeper • Help assure that the minute-recorder is recording the needed information • Make an "Issues Parking Lot" flip chart with issues to discuss later • Lead a brief meeting-process evaluation: what went well; what to improve

- Finally, clearly understand how you will conduct the meeting and deal with possible challenges. To help prepare for this *process* part of meeting planning, consider:

  ✓ If this will be the first time the group has met, draft proposed rules of group conduct for the group to revise and achieve consensus on during that first meeting; also, orientation kick-off materials should be prepared for each participant. If the rules have already been established at an earlier meeting, print a list of the meeting rules for all participants.

  ✓ Decide how you will deal with unexpected issues brought up during the meeting, but which are not on the agenda—for example, will these issues be placed on an "Issues Parking Lot" flip chart to be later discussed off-line, at the end of the meeting, or at a later meeting?

  ✓ Will you allow some potential alterations to the agenda mid-meeting, and, if so, how will this be done? For example, the group might reprioritize the agenda under your direction.

  ✓ Consider win/win approaches to group discipline, if the need arises. If discipline is negative, you will lose the participant and you will likely lose the respect of everyone else present.

  ✓ How will you work to reach consensus on issues during the meeting; also, what will you do if the group cannot reach consensus?

- **Rule Five.** *Near the beginning of the first meeting, the group should establish its own rules of conduct; these rules should be revisable by the group at future meetings upon request and by consensus.* Note that the word *consensus* implies general or harmonious agreement of the group members—they might not each agree one hundred percent, but they do agree willingly to go along with the group's rules toward achieving the group objectives. Note also that some of the group's meeting rules might need to have clearly defined consequences if they are broken. Some possible examples of group meeting rules of conduct are shown in Table 4.2.

- Although it is possible for groups to function without agreed-upon rules for meeting conduct, "Common, or at least harmonious, goals or purposes are…not criteria of groups, but of *effective* groups." (Hersey et al., 2001, 318)

- While considering meeting rules, don't forget that one of the primary objectives a leader must have is to build the group and to positively develop the individual group members. All meetings, and especially the first one, should strongly engender member participation and teamwork. Group and individual morale is critical to achieving group objectives. Overly oppressive rules can backfire. The group members need to feel welcome, needed, and honored to be present.

Rule Six. *Understand the fundamental meaning of the word* communication *and apply it well and continuously.* The leader and meeting members should use this understanding before, during, and after the meeting. The word *communication* means literally "the process of becoming like one or as one." By this definition communication cannot be one-way. The process of becoming like one involves much give-and-take, much effort to try to put oneself in the other person's shoes, or as Covey (2004, 237) says, "Seek first to understand, then to be understood." It is truly a coming together in understanding and purpose.

Before, during, and after group meetings this concept of communication, if applied diligently, will reap tremendous rewards toward having meetings be truly effective. This application of communication will aid a group to build true team spirit and unity in working toward and achieving a common goal. It is also a builder of human dignity and effective human interpersonal relations.

TABLE 2: Example Meeting Rules of Conduct (select less than 7)
• We arrive early; start, and end on time • We will work to achieve the meeting objectives • We will come prepared and stay on track • A person missing more than two of four consecutive meetings may be dropped from the group, subject to a majority vote • To meet, 75% of the group must be present • We strive to accommodate all members when scheduling group meetings • We will follow good behavior etiquette and courtesy toward each other and strive to achieve win/win results as much as possible • Meeting minutes issue within three working days • We practice Empathic Listening; Seek first to understand, then to be understood • We will not dominate discussions, but we will actively participate • We will not do or say things that discourage other members from participating • We build on each other's ideas • One person talks at a time; no side conversations • No gossip • We don't make remarks that are sexist, racist, prejudiced, sarcastic, profane, or negative to any religious, social, or other organization or person • We evaluate things, issues, and processes, not people, unless the meeting purpose is specifically to evaluate personnel for feedback and building • Ideas belong to the group, not the individual • The group speaks with one voice after decisions are made; leave united • Turn off cell phones, beepers, and alarms • Keep your cool. If you lose it, you are wrong

Rule Seven. *Show by example and teach group members how to empathically listen to each other.* Again, as in Rule 6 above, this applies before, during, and after the meeting. Empathic listening is essential to communication as described above. It must be based on sincerely caring about the other people involved; otherwise it will be perceived to be fake or manipulative. One must seek to really care about other people. "The…listener's full attention is on the content of the message *and the intention* of the speaker." (Hersey et al., 2001, 305; italics added) Listening for the speaker's feelings is just as important as listening for the facts the speaker presents. See Figure 4.

Following are some suggestions you and your group members can use to improve empathic listening.

- *Listen for the concepts and details of the message*. Make an effort to really listen precisely what the speaker is saying.
- *Listen for the feelings of the speaker*. Try to perceive the speaker's feelings, desires, and emotions an about what is being said through the way that the message is delivered—try to really put yourself in the other person's shoes and try to see things they way they see them.
- *Respond to the feelings of the speaker*. Sincerely reflect to the speaker what you recognize and understand the feelings expressed to be.
- *Watch for the speaker's verbal and nonverbal messages.* Attempt to identify the real messages the speaker may be expressing.
- *Reflect back to the speaker the concepts and feelings you are hearing*. In your own words try to share with the speaker what you think the speaker said and felt. Listen to how the speaker responds to your reflection and interact iteratively until you both understand each others messages and feelings. (Based in part on Hersey et al., 2001, 306, quoting Rogers, 1951, and on Covey, 1989)

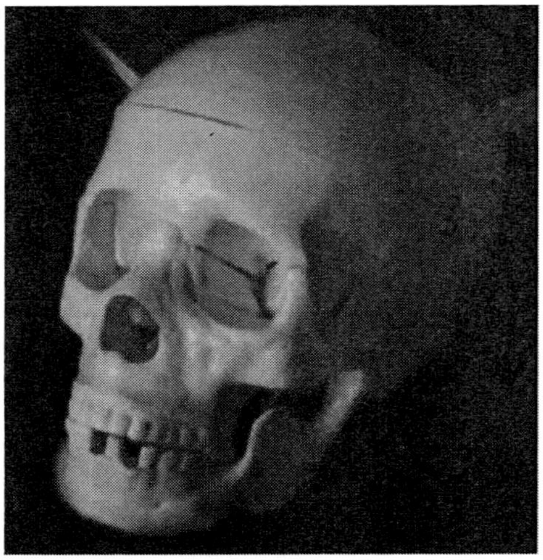

FIGURE 4. I've Been Waiting a Long Time for You to Listen to Me

The advantages of empathic listening are many. Here are a few.

- You will better understand the perspective of the other person, which will place you in a better position to cooperate,
- People will know you care about them,
- You obtain feedback,
- You gain better perspective and understanding,

- Mutual acceptance and respect are engendered,
- You will more easily achieve win/win solutions,
- You will more easily achieve synergy among the group members (Covey, 1989),
- Your group will be less likely to make costly mistakes due to misunderstandings,
- Your group will be more empowered due to better understanding and knowledge, and
- Your group will have better morale.

Many years ago I went to my manager's supervisor with a concern about how we could increase the availability of current technical literature for the engineers in our company. As I entered the supervisor's office, he shut the door behind me, sat down at a desk with me, and made me feel at ease. We had a little meeting. As I explained my concern, the supervisor took out a notebook and carefully noted my comments. He asked clarifying questions to make sure he understood. He asked for my suggestions and feelings. He summarized what he thought he heard me say and made certain that he understood correctly. Finally, he seemed genuinely appreciative that I would give him these comments and suggestions. As I left this man's office, I felt both that he had fully understood and that he cared. I also remember thinking that it really didn't matter to me anymore whether the supervisor followed my recommendations or not. Because I felt so much that he really understood me, I felt I could trust his judgment to do what was best. He had won the battle by empathic listening, whether he took any action or not.

Rule Eight. *Create a supportive and task-oriented environment during the meeting.* This rule will be accomplished as the Rules 1-7 above are followed, coupled with a dedicated effort by all meeting members to accomplish the meeting objectives. The leader and, if present, the facilitator play the crucial roles in accomplishing the benefits of this rule. They need to stay focused on the objective and care about the people! A leader who does not follow this rule will be the major cause for a miserable meeting; conversely, a leader who applies it properly will almost always have an effective meeting and one people can positively remember.

A task-oriented environment will help the meeting stay on track toward the meeting objective. Sometimes it is useful to have the team participate directly in the effort to stay on track. At the beginning of one meeting I attended, the leader handed to each member a sheet of paper with the words "We Are Off Track" in large bold letters. Anytime any group member felt the discussion was going off track, that member could hold up his sign. The result was both fun and effective.

In a supportive and task-oriented environment, meeting participants help each other, rather than hurt each other. For example, it would be inappropriate speak degradingly to or to yell at someone in front of other people. While there could be very *rare* exceptions, it is almost always not good practice to discipline someone in front of other people.

It is not appropriate to "launch missiles and throw grenades" during a meeting. Launching a verbal attack on someone in front of other people alienates them and makes a fool of you. If you do this you will lose the respect of the other meeting participants as well. If you lose your cool, you are wrong. People have been asked to participate in the meeting for a reason—or at least they have if you are following good meeting rules. Do not alienate them or they will not be able to fulfill their purpose in the meeting—they might be there in body, but they will not be there in spirit and will likely stop contributing. Thus, everyone loses, especially you!

In a supportive and task-oriented environment, meeting participants don't hold "Two meetings at the same time"—meaning that they don't carry on side conversations during a meeting. See "Don't hold two meetings at the same time." (STORY)

FIGURE 5. Create a Supportive and Task Oriented Environment

Don't come with hidden agendas to a meeting. You are out of place to do this. If you do not agree with the agenda, try to resolve this is one-on-one with the meeting leader, *well before* the meeting. Occasionally, if several of the potential meeting participants do not agree with the agenda, they could all meet in advance with the meeting leader to try to work out a compromise. In any case, when the meeting begins the time has passed to come with your own agenda.

In a supportive and task-oriented environment meeting participants sincerely phrase critiques by beginning with words such as "I wish I understood better..." or "Tell me more about how..." By doing this you might find that one of the following happens: (1) As the person explains more about their idea, they may actually convince themselves that it isn't what they initially thought, (2) As the person explains more, you might realize that it isn't what you initially thought, (3) As the person explains more, he feels so glad you are actually listening that he doesn't feel a need to pursue that idea forcefully—if you don't listen he is likely to pursue it forcefully, (4) you might find that you actually agree, or (5) you might get an new idea inspired by the other person's explanation, which idea you might not have thought of otherwise. The bottom line is to be supportive in your interactions with other people and their ideas.

Finally, in a supportive and task-oriented environment, all meeting participants should work to achieve the meeting objective(s) and not waste people's valuable time.

Rule Nine. *Close the meeting effectively.* The following hints can be very effective in closing the meeting.

- Assure that assignments and other action items are clearly defined. Use action verbs in the description of all assignments—action verbs tell what action to take and they are powerful
- Clearly decide who will do each assignment
- Clearly decide when each assignment is to be completed. If at all possible, the person who is to do an assignment should be the one to decide when that assignment will be completed
- Decide to whom the person doing an assignment will account for its status and completion—this is almost always forgotten, but can be a very helpful clarification leading to high effectiveness
- If this group is to meet in the future, have the group do the following three actions
  - ✓ Draft an objective and agenda for the next one, two, or possibly even three meetings—these are only drafts but will keep the group focused on the long-term vision of where it is going and give it a fuller sense of purpose and the need for timely performance. Note: This will be too time-consuming if built from scratch in the meeting. In advance, the meeting leader should have prepared rough drafts of these future meeting objectives and agendas for the team to revise as appropriate; otherwise this step will become disliked by the team.
  - ✓ Set the time and place of the next meeting. Do this after the objective and agenda are drafted for the next meeting(s) so that the group can identify how soon they should meet.
  - ✓ Critique the current meeting. What went well? What could be done to make it better? —This critique should not last longer than five to ten minutes or the group will not want to do it again. Keep it positive! I usually ask the group to complete the following two phrases: "I liked…" and "I wish…"

Rule Ten. *Follow up.* The meeting leader needs to build trust, accountability, and respect by following-up on assignments and action items in a caring but timely way as appropriate for each meeting member and each task to be completed. There are two objectives here:

(1) build teamwork toward accomplishing the assigned tasks and (2) build the meeting members so that they progress based on their individual situations.

Recommendations

How does one go about implementing the above rules? At first, it could seem overwhelming. It takes time and experience to implement these rules correctly. However, if meeting participants will keep the end in mind—that they want to reach their objectives in a way in which everyone wins—then they will have made the first and most important step. For many people good meeting management will be an easy and exciting step forward; for others it will be a culture shock. In all cases it is needed.

The above rules, lists, and suggestions are really tools and principles that need to be applied according to the needs of the situation and the people. Small steps at first are appropriate so that people can learn them and buy-in to the better ways. It will not take long for meeting participants to see the benefits and the need to change. To run too fast into full implementation of these rules could cause rejection due to misunderstanding and the natural tendency by many of us to hold to the *status quo*. Care deeply about your people, but persistently move forward toward helping them to have more effective meetings. Most people will welcome the needed changes.

Conclusion

Conducting effective meetings takes a concentrated effort, but the return on the effort invested will pay big dividends both during the meeting and for the long term. Objectives will be clearer and will be achieved. Morale will be higher, as will productivity. There is really little or nothing to lose by doing it right, but a great deal to gain. Meeting members will want to come back again, rather than dreading the words, "Another meeting."

References

Covey, Stephen R., *The 7 Habits of Highly Effective People—Restoring the Character Ethic,* Simon & Schuster Inc. (1989).

Dinsmore, Paul C., *Human Factors in Project Management,* AMACOM (1990).

Dressler, C. 1996. Cited by J. P. Lewis, *Team-Based Project Management,* AMACOM (1998). Cited also by Karl A. Smith, *Project Management and Teamwork,* McGraw-Hill (2000).

Gido, Jack, and James P. Clements, *Successful Project Management,* South-Western College Publishing (1999).

Hersey, Paul, Kenneth H. Blanchard, and Dewey E. Johnson, *Management of Organizational Behavior—Leading Human Resources,* 8th ed., Prentice Hall (2001).

Meredith, Jack R., and Samuel J. Mantel, Jr., *Project Management—A Managerial Approach,* John Wiley & Sons, Inc. (2000).

Rogers, Carl, *Client-Centered Therapy: Its Current Practice, Implications, and Theory,* Houghton Mifflin (1951). Cited by Hersey, Paul, Kenneth H. Blanchard, and Dewey E. Johnson, *Management of Organizational Behavior—Leading Human Resources,* 8th ed., Prentice Hall (2001).

The Importance of Excellent Definition of Project Objectives

Abstract ID: 481746
Paper # 330e

Dr. Eldon R. Larsen
Marshall University

Effective project objectives are vital for successful project management. They tell us what we want the end service, result, or product of a project to be. Surprisingly, most project management texts go little beyond acknowledging the importance of good objectives. The *"How to"* is either missing or given only very brief discussion. This paper presents some basic principles and tools helpful in setting effective objectives. The tools help address questions around project scope, decision-making trade-offs, and tolerance between what is ideal and what is failure. These same tools aid in communication between the project team and other key players so that they all understand the intended project outcome.

Introduction

Clear, specific, effective objectives give us the best picture of a project's desired destination. "Objectives are the aims or end points toward which activity is directed." (Babcock, 1991) Someone once said, "I always wanted to be successful, but I guess I should have been more specific. I should have defined success."

When working on projects, what do the project objectives really mean? Do we really understand them? Do other people understand them the way we do? How will we know when the objectives have been achieved? Is there any room for variation if we have trouble reaching the ideal? Have key people agreed upon the objectives? What constraints are we facing? Which constraints are most important and which least important?

These and other questions are the topic of this paper as we investigate one of the most important parts of starting projects right—how to develop effective project objectives.

Tool 1: Make Objectives SMART

One very helpful tool for developing effective project objectives is the acronym SMART, described in **Figure 1**. Project objectives need to be specific, measurable, agreed-upon, realistic, and time-oriented. A weakness in any one of these areas can be either like a hidden time bomb ticking away and ready to blow up at any time, or like a cancerous problem that slowly grows until it either kills the project or requires it to have major surgery and expensive life-saving resuscitation. Of the many variations on what the letters in the acronym SMART stand for, the author believes the one presented in **Figure 1** to be most useful. It is based on Smith (1991), but the author has expanded the definitions for clarification.

S	**S**pecific:	Targeted, focused, clear, concrete, understandable, documented, not abstract, not general, not overly complex
M	**M**easurable:	You can tell when you've reached it, targets are clear, tangible, verifiable, and quantified if at all possible
A	**A**greed-Upon:	Everyone involved is aligned; owner, senior management, project leader, project team, and other key people have approved and committed, preferably in writing
R	**R**ealistic:	Challenging, but attainable; not impossible to achieve nor viewed that way; consistent with resource and time constraints; not too rigid to accommodate changing priorities
T	**T**ime-Oriented:	Clear time frame for project completion

Figure 1. Objectives Should Be SMART

Specific. A project objective that is *specific* will be targeted, focused, clear, concrete, understandable, documented, not abstract, not general, and not overly complex. What we are trying to do with the objective statement is to be very good artists and to paint a very clear picture for everyone involved with the project. That artistic picture should not be like the impressionistic artistic style such that the words appear pleasing from a distance, but when you get close up there is not enough detail to understand what is intended. Instead the style needs to be more along the realistic artistic style, where both far-away and close-up examinations provide good understanding.

Part of being specific is to put the objective in writing. Sometimes for small projects we feel this is not worth the trouble. However, the act of putting an objective in writing forces us to clarify it. If you can't write it down, you don't understand it.

Measurable. The project objectives also need to be *measurable*. How will we know when we have achieved the desired outcome? We need to be able to measure this in some specific way. The objectives need to be clear, tangible, verifiable, and quantified if at all possible. "You can't prove a project's success if there aren't any measures for it." (Frye and Mason, 2001)

I once worked on a research project for approximately four years with a very fine team of people. We met regularly and were very familiar with what was happening on the project. Yet, after four years I started hearing comments in the team meetings that in effect said, "Are we finished now?" Some members thought the answer was "Yes." Others thought "No." Upon examination of the objectives, we could not tell. The objectives were neither clear enough nor measurable enough to allow us to make that decision—after four years!

In making objectives measurable, one key is to decide both *what* you want to measure and *how* you will measure it. The *what* and *how* both need to be specific enough so that everyone knows when they are met. Sometimes the measurable characteristics of the project objectives are called metrics. Regardless of the terminology, we need to answer the *what* and *how* measurement questions.

For example, one of the objectives of a project might be to build a sailboat that is thirty-six feet long. One part of the *what* might be the dimension thirty-six feet. *How* we can measure the objective might be with a tape measure. This example is fairly easy.

Another example might be that we have an objective for the people at our family reunion to enjoy the food provided. The *what* is a little harder to quantify here, to enjoy the food. The more difficult question is *how* to measure that enjoyment. We could have an end-of-reunion critique meeting in which the food is discussed; we could give everyone a written survey; or, we might come up with another more creative method of measurement. The point is that if we do not define how we will measure the objective, there is a good chance it will not be measured and we will not be able to determine if the objective has been achieved.

Let's look at one last example of a measurable objective. A hotel manager might have as an objective, "To increase the hotel owner's satisfaction with my management performance." Although this is a very nice desire, the objective is neither very specific nor very measurable. First, the hotel manager needs to define better what would affect the owner's satisfaction. Second, the manager needs to define better what the word *increase* means, as this could be anything from a very small change to a very large one. What does the owner want? What can the manager provide? As an example, perhaps a more specific and measurable objective might be, "To increase the average hotel room occupancy from the past value of eighty-five percent to a new value of at least ninety-five percent during the months of June through August of this year." This revised objective is much more measurable and specific.

Three cautions are important when making objectives measurable. First, watch out for common words that sound desirable, but are not specific and not measurable. A few examples are the words *good, excellent, pretty, high quality, long lasting, preferred,* and *successful*. To correct the problem with these words, simply express in quantifiable terms what you really mean. For example, the phrase "We will provide excellent help-desk service," might be changed to something like, "We will answer the telephone for ninety-five percent of all help-desk calls within 20 seconds and we will provide solutions acceptable, according to the customer, to at least ninety percent of the problems within one working day."

The second caution is to be careful not to skew your objectives just so you can easily measure them. This can lead to disaster just as badly as can the original immeasurable objectives. The challenge is to re-express your real objectives in a way that they are both clear and measurable as much as is realistically possible.

The third caution is to be very careful that what you choose to measure, and how you measure it, do not cause unwanted performance side effects. Sometimes while driving desired behaviors, we inadvertently drive new, unwanted behaviors. There is an old saying that you get what you measure.

Agreed-upon. The project objectives need to be *agreed-upon*. This means that everyone involved with the project is aligned and that, preferably in writing, the project owner, senior management, project leader, and other key people have approved and committed themselves to the project objectives.

Some project owners give the project team a set of objectives and simply expect them to go to work and achieve them. While this can work in some cases, experience shows that participant buy-in is usually very important for effective organizations. This is true as well for project organizations and teams.

By allowing a project manager, and where practical the project team, to have a chance to work with the project owner on the definition of the objectives, a far better set of project objectives is probable, and participant buy-in is more likely and much higher. In the long run, this extra effort pays big dividends and is more likely to produce win/win situations for the owner and the project participants.

When an owner gives a team project objectives with which they might not fully agree, and without any way to iteratively improve those objectives through two-way communication, that owner unintentionally is also often asking for ineffective results, inefficiencies due to poor morale or poor understanding, and possibly even project failure.

If used properly, the tools and concepts presented in this paper can actually be of great help in improving communication between the key people involved with a project and go a long way to facilitating agreement upon objectives.

Realistic. Project objectives should be *realistic*, meaning that they are challenging but attainable. They should not be impossible to achieve nor be perceived that way. They should be consistent with resource and time constraints. And, order to accommodate changing priorities when essential, they should not be too rigid. "Generally, the more difficult the goal, the higher the level of performance. But…if and when a goal is perceived as so difficult that it's virtually impossible to attain, the result is often frustration rather than achievement." (Gibson, et al., 1997)

One caution is worth mentioning with respect to making objectives challenging: Occasionally a project owner will allow conflict to exist within the project objectives, perhaps even intentionally, thinking that higher achievement will result. This is normally not a positive motivation for teams. Gibson, et al. (1997) report that a recent study "examined conflict as a goal-setting variable. Both a laboratory experiment and a field study of college professors suggested that conflicting goals can lead to decreases in performance…This research points out the need to pay attention to the total situation."

In summary, project team members will be better motivated both as individuals and as team members when objectives challenge them in a positive way, but do not overwhelm them due to either real or perceived lack of attainability.

Time-oriented. Finally, project objectives need to be *time-oriented*, meaning that there is a clear time frame for project completion. Sometimes, particularly in research and development, teams are given project objectives without any idea as to what the time-limit expectations are. Although there are situations where time limits are difficult to define, even in those cases an approximate time *range* in which a project needs to be completed should be describable. For example, a research project outcome and endpoint might be uncertain, but funding cannot continue indefinitely. In such a case, we might have to say something like "This project is to be completed within the next two to three years." Although some people might feel uncomfortable with this flexible time range, it is much better than saying nothing! And when there is high uncertainty, an ending time *range* in not inappropriate. That range can be narrowed down as time progresses and the project develops. We will talk more about *time* as part of our discussion of the triple constraints, below.

Tool 2: Make Full Use of the Triple-Constraints

Another important concept and tool is that of the triple constraints, illustrated in **Figure 2**. A constraint is something that imposes a limit or boundary. In project management, we normally speak of three major constraint criteria: the triple constraints. Although there are some variations of this concept, the author has found the one shown to be generally very useful, the three general constraints on the project being *time*, *resources*, and *performance* criteria. Let's examine each one.

Time Constraint. The *time* constraint, already introduced briefly above, is relatively easy to understand; at least that could be our impression. However, perhaps it is this impression of easiness that causes people occasionally to overlook adequately describing it when they write project objectives. The time constraint is very important. It basically involves identifying when the project is to be completed, at least roughly. Associated with the time constraint is how long it will take to do the project—the project duration. It is not the purpose of this paper to go into the many tools useful for planning a project and estimating project durations. The point here is that it must be done.

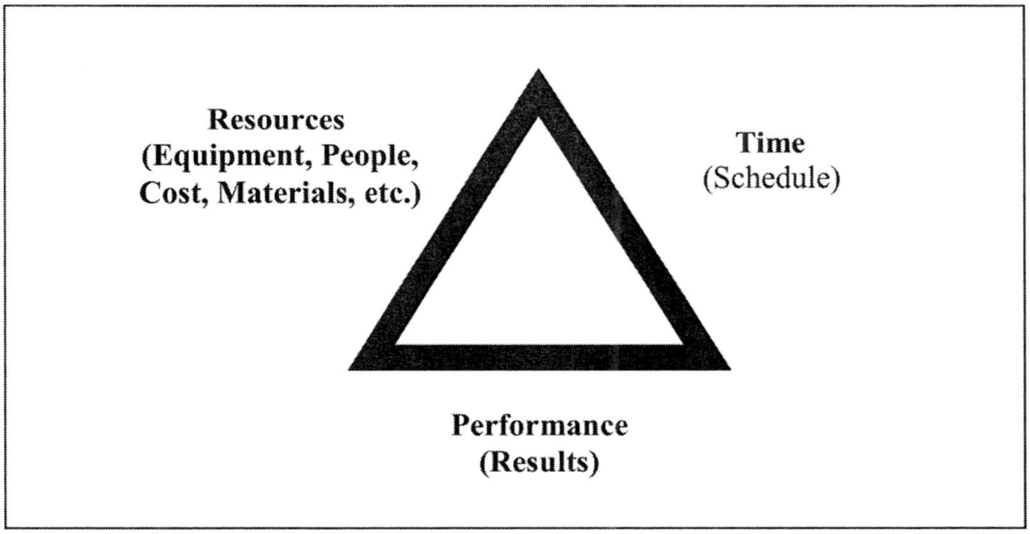

Figure 2. The Triple Constraints

It is not uncommon for the owner of a project to give a deadline for a project without due consideration as to whether the deadline is reasonable. Normally, to determine an initial, realistic time limit estimate, at least some high level planning of the entire project is needed. This initial time estimate can be refined later, and should be, but setting time limits without at least high-level overall project planning is not only unwise and likely to cause much frustration, it can also be disastrous to the project.

There are cases where a deadline is fixed due to reasons beyond the control of either the owner or the project team. For example, legal or regulatory implications, market competitions, or business and financial constraints might dictate a deadline. When this is the case, it is still very important to do at least a high-level plan to see if the deadline has a realistic chance of being achieved. Not doing this can cause a large amount of wasted money, effort, and time.

Resources Constraint. The *resources* constraint is a little more complex. Sometimes it is simply stated as cost or budget, but in reality it includes also items like people, equipment, materials, facilities, etc. When a project owner and/or project manager focuses only on budget as the resource constraint, the other resources of people, equipment, materials, etc., can be neglected until problems have had a chance to develop to the point that project execution is hampered. The message here is that all the project resources and their interactions need to be considered as a constraint on the project.

Performance Constraint. The third general constraint is *performance*, or results. It includes many possible specifications and is sometimes, but not consistently, referred to as project *scope*. Sometimes performance is defined as the project technical objectives, but in reality it can include much more, or, depending on the project, it could have nothing to do with anything technical at all. Examples of some of the items that might be considered for inclusion in the performance are given in **Figure 3**. Care must be taken not to neglect these possible aspects of performance when writing the project objectives. Some of the items in **Figure 3** were taken from Bitz (1991).

- Deliverables
- Quality
- Production rate
- Product specifications
- Customer satisfaction
- Stakeholder satisfaction
- Labor relations
- Safety
- Health
- Environment
- Work processes
- Quantity
- Reliability
- Operability
- Modularity
- Flexibility
- Maintainability
- Community relations
- Corporate image
- Product cost

Figure 3. Some Example *Performance* Constraints

Interrelation of the Triple Constraints. Note that the triple constraints are not independent of each other. A needed change in the time constraint can affect resources. A change in the resources can affect our ability to meet a time constraint, or cause a need to change that time constraint. Changes in performance criteria due to new knowledge or new situations can cause a need to adjust both the resource constraint and the time constraint. Therefore, use caution when modifying constraints to assure that the effects on the other constraints are well understood and accounted for.

Triple Constraints Are Project Drivers. There is a general understanding of the triple constraints by many project managers, but one of the most important uses of the triple constraints is often either unrealized or neglected—the concept of the triple constraints being the project drivers. By *drivers* we mean: Which one of the triple constraints (time, resources, or performance) is most important, i.e., the *primary driver;* which one is least important, i.e., the *weak driver*; and which one is in the middle, i.e., the *middle driver*? (Dobson, 1993) Do we really need to care about this? Aren't time, resources, and performance all important? Do we really have any choice? The answer to all of these questions is, "Yes."

The truly vital question to understand when discussing the importance of the drivers is, "Why do we care?" We care because at some point in the project we are going to have to make vitally important decisions with respect to the project, and the correct decisions are going to depend on the driver priority or importance relationship.

All three of the triple constraints are important for all projects, but they are *not* all three *equally* important in any project! The trick is to determine which constraint is most important, which least important, and which is in the middle for the particular project concerned. How do we do this?

Well, we could ask the project owner which driver is most important, time, resources, or performance. In response the owner is likely to say something to the effect, "What do you mean by asking me this? They are all essential and I expect you to give me them all. Now go back to work." This approach will usually fail. The problem is that even the project owner usually does not think in these terms. You will have to help him.

One way to do this is by asking the project owner a variety of trade-off questions. It is best first to patiently explain to the owner what you mean by the triple constraints. Tell the owner that you know that each of the constraints is very important. Then, tell the owner that because you really want to manage the project so that he or she is satisfied, there will come times during the project that you will have to make important decisions, and you want to do your best to make the right choices. Then, give the owner some example general trade-off questions that you might be faced with. Don't get too detailed; stay general—you probably don't really know at this point what problems you will run into and what decisions you might face. Right now we are talking hypothetical. What you are trying to do is to weigh each possible pair in the triple constraints—time vs. resources, time vs. performance, and resources vs. performance.

For example, you might say something to the effect, "Although we don't want this to happen, suppose as the project progresses we realize that in order to finish by the project due date, we must hire some people to work overtime, thus increasing the cost a little, and causing us to go over budget a little. Would you rather have us do that or would you rather we stay on budget and finish the project late by a week or two?" The project owner normally will be able to give you an answer to this question that gives you a feeling for whether resources or time is more important—he might even explain the reason, which would be very helpful insight that you did not have earlier.

Then, you could ask a trade-off question between the performance and the time; then, the performance and the resources, and so forth. You should ask enough different questions so that you get a clear feeling for the proper order of the triple constraints: which is the primary driver, the middle driver, and the weak driver.

Finally, state clearly to the owner what you have perceived to be the constraint priorities and get verification that the owner is in agreement. Putting it in writing would be best.

Driver Cautions. There are several cautions to be aware of when determining the triple constraint drivers.

First, there are things that can develop during the course of a project that cause the order of the drivers to change. Watch out for this possibility. Frequent communication with the owner and the other key project personnel is vital to help you to constantly be aware of which constraint is most important, etc. A change in driver priorities should only occur when absolutely necessary, and then only with agreement among the key decision makers on the project, such as the owner, the project manager, and other key leaders. Such a change should always be rapidly communicated to all people involved with the project.

Second, you should not purposely plan the project with the intent that the order of the drivers will change. For example, at the beginning of the project you don't say that for the first half of the project we will have time be the primary driver and then for the second half we will have performance be the primary driver. If you and the owner have agreed that performance will be the primary driver at the end of the project, then you plan the entire project from the beginning as if performance is always the primary driver. Then, if the primary driver must change, it changes, and you re-plan accordingly.

Third, you might find that two of the three drivers are both extremely important, with the third being less important. Does this mean that you should declare a tie and have two primary drivers and one weak driver, with no middle driver? The answer is "No!" Ask more trade-off questions between the two drivers that you think are tied. Even though one of them is to be designated a middle driver, this does not mean that it cannot be a very, very close second to the primary driver. It can. But when problems arise in the project and decisions have to be made, you must know beforehand which driver is on top.

Finally, as already discussed above, the triple constraints, and thus the drivers, are not independent of each other. Always be aware of how a change in one affects the others.

Tool 3: Project Tolerance Diagrams

Let's suppose the owner of your project specifies that the project is to be completed by July 4, 2003. Our impulse might be to treat this as the deadline and do everything in our ability to reach it, even if it means hiring people on overtime or cutting a few corners in performance to do it. On the other hand, we might tell ourselves that the owner just pulled July 4 "Out of the air," so to speak, as the date that he would like to have, but it is not essential and that sometime soon after July 4 should be all right—it is ideal but not really that essential. Which assumption is correct? Well, it could be either one. Do we really need to know this? Absolutely! The answer could mean the difference between project failure or not.

A tool to help us with this problem is called the *Tolerance Diagram*, shown in **Figure 4**.

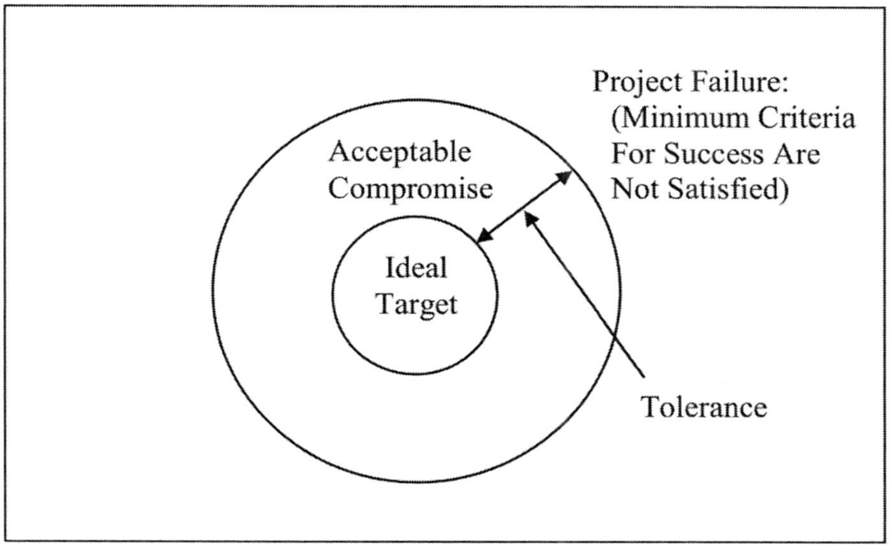

Figure 4. A Simple Tolerance Diagram

Three zones exist on the Tolerance Diagram in **Figure 4**: The ideal target zone, the acceptable compromise zone, and the failure zone. The ideal target zone describes what we would really like to have, if at all possible within the constraints on the project. The acceptable compromise zone is what we must have, even if we can't have the ideal. If we can't have at least the acceptable compromise, then we move into the failure zone and the project must terminate, or another alternative must be determined, including the possibility of a different or a significantly re-defined project. We call **Figure 4** a Tolerance Diagram because it illustrates how much tolerance or flexibility there is between our ideal desires and what might be considered failure. The primary driver will tend to have the least flexibility or tolerance, while the weak driver will tend to have the greatest flexibility or tolerance.

Let's look at an example. Suppose our only family car has broken down and this time it really is just not worth it to get the car fixed—we need to purchase a replacement. We have certain limitations or constraints, one of which is the money to purchase a car. It would of course be nice to have a new red Ferrari, but the cost is out of the question—it would cause failure in our credit rating within a couple of months. What we need is car that will reliably start and will get us to work each day—this we must have. After some financial figuring and some emotional stretching, we decide that we would like to spend no more than $5,000 on the car—this is our ideal target, a cost less than $5,000. We also decide that if we really can't find a car for a cost less than $5,000 that meets some minimum reliability standard (reliability is a performance constraint), we would be willing to go as high as $7,000, but absolutely no higher. Our acceptable compromise for the money constraint is therefore a cost that is between $5,000 and $7,000. Failure is to have a cost greater than $7,000.

People familiar with various problem-solving techniques will note some similarity between the one-sided Tolerance Diagram described above and the suggestion by Kepner and Tregoe (1981) that there are *wants* and *musts* that need to be identified when trying to evaluate alternative solutions to problems. Although Kepner and Tregoe's wants and musts are not exactly the same respectively as the ideal target and the acceptable compromise zones of the Tolerance Diagram, the detailed analytical approach they use to weigh various alternatives could conceptually be applied to the Tolerance Diagram, and interested readers are referred to their work.

Figure 4 is a *one-sided* Tolerance Diagram, meaning that going from best to worst is one-directional. Rather than the simple one-sided Tolerance Diagram, some project specifications will require a *two-sided* Tolerance Diagram, as illustrated in **Figure 5**. In this case the ideal target is a range in the middle of a project specification or constraint. On either side of this ideal target range are ranges deviating to one side or the other, which are still acceptable, but not ideal, giving therefore two acceptable compromises. Going outside of these two acceptable compromise ranges causes the project to move into unacceptable ranges, which again we call failure.

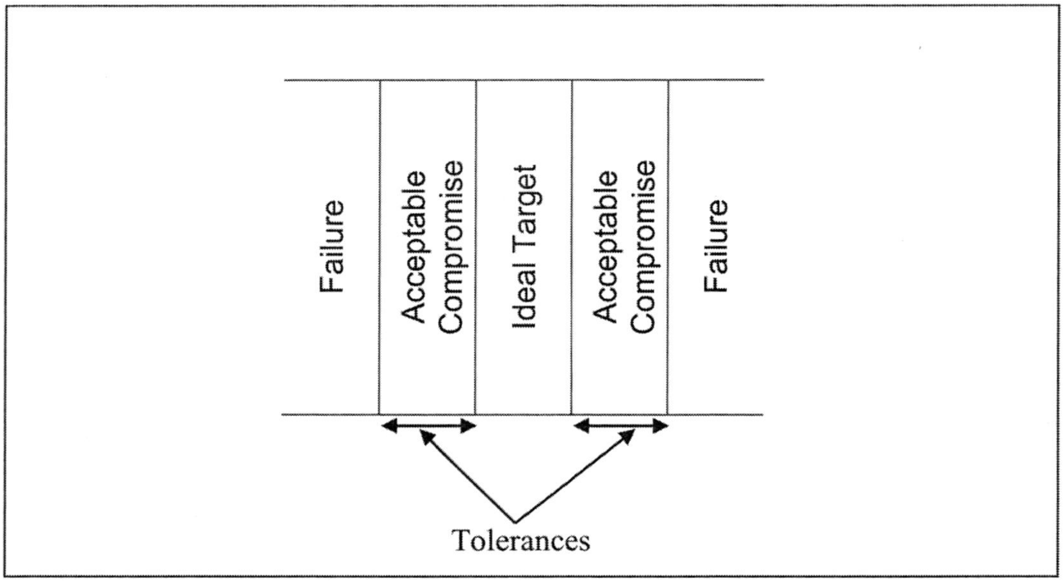

Figure 5. Two-Sided Tolerance Diagram

Let's look at an example of a two-sided Tolerance Diagram. Suppose that we are trying to make a new polyethylene polymer product that meets a certain density specification. We might want to have the polymer density be 0.950 g/cm^3. It will not be possible consistently to make that polymer with a density of exactly 0.950 g/cm^3, so instead we might specify that our ideal target range is 0.948 g/cm^3 to 0.952 g/cm^3. Although this is the ideal target, if we really must we could accept as a compromise polymer as low in density as 0.946 g/cm^3 or as high as 0.954 g/cm^3. Therefore, one acceptable compromise range would be 0.946 g/cm^3 to 0.948 g/cm^3 and the other would be 0.952 g/cm^3 to 0.954 g/cm^3. Any density less than 0.946 g/cm^3 or greater than 0.954 g/cm^3 would be failure.

Care should be taken when developing project objective specifications and identifying constraints to understand whether we need a one-sided or a two-sided Tolerance Diagram. Many projects will use both, some specifications or constraints using one-sided Tolerance Diagrams and the other specifications or constraints using two-sided Tolerance Diagrams.

Generally there should be a minimum of three Tolerance Diagrams for every project: one for each of the triple constraints of time, resources, and performance. However, usually you will need more than three. This is because the resources constraint and the performance constraint each have several aspects to them.

Looking at performance for example, we might have one Tolerance Diagram for some quality specification, another Tolerance Diagram for meeting some safety objectives, and yet another Tolerance Diagram for a new product production rate.

The author has seen an approach similar to Tolerance Diagrams receive very favorable reviews by the project team on a very large successful industrial capital product. The team reported in an end-of-project review that it had never before seen better-defined project objectives.

One of the arguments sometimes given against using the tolerance diagram concept is that perhaps people will not work as hard if you tell them the acceptable compromise in addition to the ideal target. This argument is weak for the following reasons.

First, if people have worked for you before, they have already figured out if you have habitually told them solely the ideal target, almost always to compromise it later on. In this case it is better to tell them the acceptable compromise up front rather than for them to make one up in their minds and be continually second guessing you.

Second, if you have a proper reward structure that rewards project teams for attaining the ideal objective as often as possible, they will respond to the rewards.

Third, people learn to trust you better when they know you are fully communicating with them and telling them the whole story—telling them both the ideal target and what is an acceptable compromise, if such a compromise is needed. This way the team and the owner can establish a more truly honest working relationship.

And fourth, it is much better for a team to know up front that an objective is either the ideal target or else the least you can accept, rather then for them to second guess this and have a 50% chance of being wrong, causing everyone a lot of frustration, wasted time and effort.

Some Additional Considerations

When developing project objectives there are additional considerations that can be important. **Figure 6** gives some of these considerations.

There can be times when project objectives have been given to the project leader and team, without due consideration to the material presented. In these situations the objectives will frequently be lacking in quality and need to be refined or re-done in order to be effective. There are times, in other words, when the project objectives really need to be re-negotiated.

- Verify that you are solving the right problem
- Talk to experts and those who might do the work
- Consider more than the first impression of what would be ideal for each of the triple constraints
- Are there better technologies or more cost-effective ones?
- What is changing in the market or new technology environment?
- How will key personnel affect this project?
- Document all the assumptions you are making
- Explore possible alternative solutions or objectives
- Prepare a draft set of objective statements, then test this draft with key people
- Refine the draft objectives and again review them with the key people
- Make sure that you have alignment of business and technical objectives
- Make sure that success metrics have been defined
- Assure that project implementation expectations are clear with all parties involved
- Once the objectives are finalized, constantly remind everyone involved of those objectives
- The longer the project duration, the more likely the project objectives might change or evolve—be alert for this possibility and manage the changes carefully through full communication

Figure 6. Some Additional Considerations for Developing Project Objectives

As Knutson and Bitz (1991) have stated well, this negotiation step "is a result of one of two sets of circumstances: (1) You approach the client and senior management for the first time in order to negotiate a schedule deadline and a budget because these were not established when the project was initiated, or, more commonly, (2) you realize that there is an incompatibility between the preliminary plan and the project objectives that cannot be eliminated by negotiating estimate revisions…In the latter case, the negotiations are more complex…There are four alternatives for management to consider:

1. Alter the schedule and cost objectives
2. Develop a reduced set of technical objectives (referred to as "Performance" objectives in this paper)
3. Implement a combination of the first two alternatives
4. Cancel the project

Regardless of the alternative you recommend and senior management decides on, there are only two acceptable outcomes to this process: (1) a plan acceptable to you, the project team, the client, and senior management or (2) cancellation of the project. There is one other possibility, however—one that we do not encourage: taking on the project on a best-efforts basis, without changing the project objectives."

If the procedures and tools suggested in this paper are followed and used, the above difficulties caused by poor project objective definition will be minimized or nearly eliminated altogether.

Once we have well defined, written goals or objectives, they need to be published for all people involved with the project. As Rees (1991) has said, "Publish the goals. Make them visible for all to see. Do not expect people to remember them if they are not discussed and referred to often. Do not bury them in occasional memos and documents. Instead, post them in meeting rooms on flip charts or posters; use stickers or other visual reminders to keep the goals in front of everyone. Refer to the goals often in memos, presentations, and meetings. Whenever possible, use the goals to guide a decision."

Finally, even after all the above work to do objective development right, to varying degrees the project objectives need to remain dynamic. How much the objectives are dynamic will depend on many things, and each project is unique. But the project objectives "should be reevaluated as the environment and opportunities change." (Rue and Byars, 2000) **Figure 7** gives some example factors that have the potential to cause objectives to change significantly.

- Market competition
- Significant changes in the national or corporate economy
- New senior management
- New discoveries
- New regulations
- Unknown technical difficulties in first-time scale-up to commercial production

Figure 7: Example Factors Potentially Changing Project Objectives Significantly

One important caution is appropriate here: When project objectives change very significantly, it is often much better to declare the old project dead and to officially start a new project with the new objectives. This is because the old baggage of plans and expectations from the original project can hinder the revised project to the point of making it inefficient and ineffective. It is very hard to change the direction of a speeding train locomotive.

Summary

In summary, defining good project objectives as early as possible is one of the most important project development steps to occur. Presented in this paper are a number of important tools and concepts to help project owners and project managers with this extremely important responsibility. Particularly we have discussed in detail SMART objectives, the triple constraints, the concept of prioritizing the project drivers, and the clarification that comes through use of Tolerance Diagrams.

All of these tools and concepts both help to facilitate and require much better multi-way communication between the project owner, project manager, project team members, and other key project personnel. The resulting improvement in project objectives definition will significantly raise the potential for project success.

References

Babcock, Daniel L., *Managing Engineering and Technology—An Introduction to Management for Engineers*, Prentice Hall (1991).

Bitz, Ira, *Senior Project Management*, Course Syllabus prepared for The American Management Association, revision number 3 (1991).

Dobson, Michael, from discussion at a Fred Pryor Seminar entitled *Project Management* (1993).

Frye, Chris and Holly Mason, "Project Management Simplified," Available from Internet: <http://www.webtechniques.com/archives/2001/09/frye/> (cited March 15, 2002)

Gibson, James L., John M. Ivancevich, and James H. Donnelly, Jr., *Organizations—Behavior, Structure, Processes*, Irwin/McGraw-Hill (1997).

Kepner, Charles H. and Benjamin B. Tregoe, *The New Rational Manager*, Princeton Research Press (1981).

Knutson, Joan and Ira Bitz, *Project Management—How to Plan and Manage Successful Projects*, AMACOM (1991).

Rees, Fran, *How to Lead Work Teams: Facilitation Skills*, Pfeiffer & Company (1991).

Rue, Leslie W. and Lloyd L. Byars, *Management—Skills and Application*, 9th ed., Irwin/McGraw-Hill (2000).

Smith, Larry A., *Basic Project Management*, from course materials reprinted with permission for use in American Management Association seminars (1991).

Overview of Project Planning

Abstract ID: 481747
Paper # 330f
Dr. Eldon R. Larsen
Marshall University

Effective, excellent project planning is one of the vital three keys to successful project management, the other two vital keys being the importance of people and the importance of excellence in defining the project objectives.

When we say *planning* we are including *re-planning* as well. Re-planning is just a word we use to indicate the fact that the nature of planning should be ongoing continually throughout the life of the project. In fact, one very fine project manager I once met said that re-planning is really synonymous with controlling the project. Although people might say such a statement is over-simplified, in many ways this is a good way to think about the need to continually plan and re-plan.

To further illustrate an important point about planning, I use the following story. During the Persian Gulf War in the early 1990s the United States used some new, high-tech weaponry. Yet that alone was not enough to win the war, and it had only part to do with how the war was executed. In fact, many people recognized that the execution of the war was being successful in large measure due to some excellent leadership and decision being made by those in charge.

One day during the Persian Gulf War, at a news conference, a reporter asked the following question, recognizing how difficult it must be to plan and conduct a war when the enemy was continually moving around, hiding in difficult terrain, and moving scud missiles around on the back of trucks. The question was something to the effect: "General Schwarzkopf, how do you go about executing this war when the enemy is continually doing things that are not in your plan?" General Norman Schwarzkopf, quoting another famous general, Dwight D. Eisenhower, from World War Two, said in response that "Plans are worthless." Then, following a pause, continuing to quote, he said, "Planning is essential."

What did Generals Schwarzkopf and Eisenhower mean when they said that plans are worthless, planning is essential? In essence what they were saying is that plans are not sacred and they are not fixed! Many events and changes occur, new knowledge is learned, situations change, changes in leadership occur, competitors make unexpected decisions or announcements, people issues arise, and so on and so on and so on, all combining to make it completely unrealistic to think that somehow a plan can remain fixed! And yet there are a very large number of managers today who do not seem to understand these realities.

It is not *the plan* that adds value in project management! Rather, it is the planning *process* that adds value! It is through the *process of planning* that we have to think about all the various aspects of the project; we have to think about the various options and trade-offs; we have to think about what competitors might be doing; we have to think about alternative courses of action and the associated consequences; and we have to keep doing this throughout the entire project!

For a manager or anyone else to expect a project team to make a plan and then to religiously stick to it exactly with no adjustments is project tyranny. A manager who does this will cause his people to hate planning. Plans will become like strait-jackets, or like a noose around people's necks. This does not mean that the project objectives change or are flexible, although in some cases they might have to be, but rather it means that the way we achieve those objectives will continually have to be adjusted, modified, updated, and revised. Planning and re-planning are a way of life for project managers if they want to be successful. In fact, without re-planning, it is very likely that the project objectives will not be achieved! The bottom line to remember is this: In order to achieve project objectives, it is the continual *process* of planning that is essential and adds value, not the plan itself!

Throughout this presentation, we will be discussing many very important project management planning tools to help us in the process of project planning. These tools are very important and useful. Still it is not the tools themselves that are sacred, but the value of the tools comes through the process of using these tools. Use them appropriately.

The project planning tools in this presentation follow one another in a logical pattern. This pattern might be thought of as a cycle of project planning. The cycle is iterative.

When a project is first being planned, it might be necessary to iterate through the project planning cycle several times in order to get the initial plans and project objectives to be internally consistent with each other. During this initial iterative process through the planning cycle, an organization might find that it must make some adjustments to the original project objectives. This might happen when the planning cycle reveals better estimates for the project duration, project costs and other resource requirements, and the project performance expectations, and those estimates are not compatible with the project objectives. As adjustments are made in the objectives and the consequences of those adjustments are analyzed in the project planning cycle, an internally consistent plan and objectives can be achieved, or the project might need to be cancelled or significantly re-thought.

The project planning cycle generally consists of the following major steps:

→ Project charter development and elaboration
  - Project objectives definition
  - Identification of critical success factors
  - Definition of project standards
  - Identification of project leader and team
  - Elaboration of project deliverables
  - Other charter concerns

→ WBS development and elaboration

→ Network diagram development and elaboration

→ Estimates of resources, including costs, and durations for project tasks

    - Note: Once this information is developed, a check for best-logic-consistency should be made in the network diagram

→ Development of resource loading information (resource histograms)

    - Note: Once this information is developed, a check for best-logic-consistency should be made in the network diagram and the succeeding estimates step

→ Development of project cash flow, cumulative cash flow, and overall budget

→ Development and elaboration of plans for managing procurement, quality, risks, communications, scope, integration, human resources, time, and costs

→ Development and elaboration of contingency plans

    - During and following development and elaboration of contingency plans, it is best to check for best-logic-consistency starting back from the WBS development through this step involving contingency plans, and all the steps in between (WBS, network, estimates, resource loading, cash flows and budget, and the other management process plans just before this contingency planning step)

→ Cycle back to check for internal consistency with the project charter

The above steps in the project planning cycle are shown graphically in Figure 1. After the initial iterations through this cycle which take place during the latter part of project initiation, the project team, its leader, and all key project stakeholders should have a good idea of the realistic expectations for the project. This does not mean that all will go as initially planned—it will not—but the team will now be prepared the best it can to move forward with project execution, with a much better chance of having a happy and successful project closeout for all parties involved.

Yet, the use of the project planning cycle has really only begun when project execution starts. The project manager must regularly check and monitor all calculation, assumptions, estimations, and so forth in the planning cycle. This is the heart of re-planning and the ability to control the final outcome of the project.

Summary

The project planning cycle contains a series of steps that build one upon another. Each step in the cycle is important and essential, and should not be skipped except in the most unusual circumstances, such as in a very small project that possibly does not merit a full use of these tools and processes. A wise project manager will give sincere consideration to the concepts discussed in this presentation. Doing so will help him to achieve true excellence in this, the third vital key to successful project management: Excellence in project planning and re-planning.

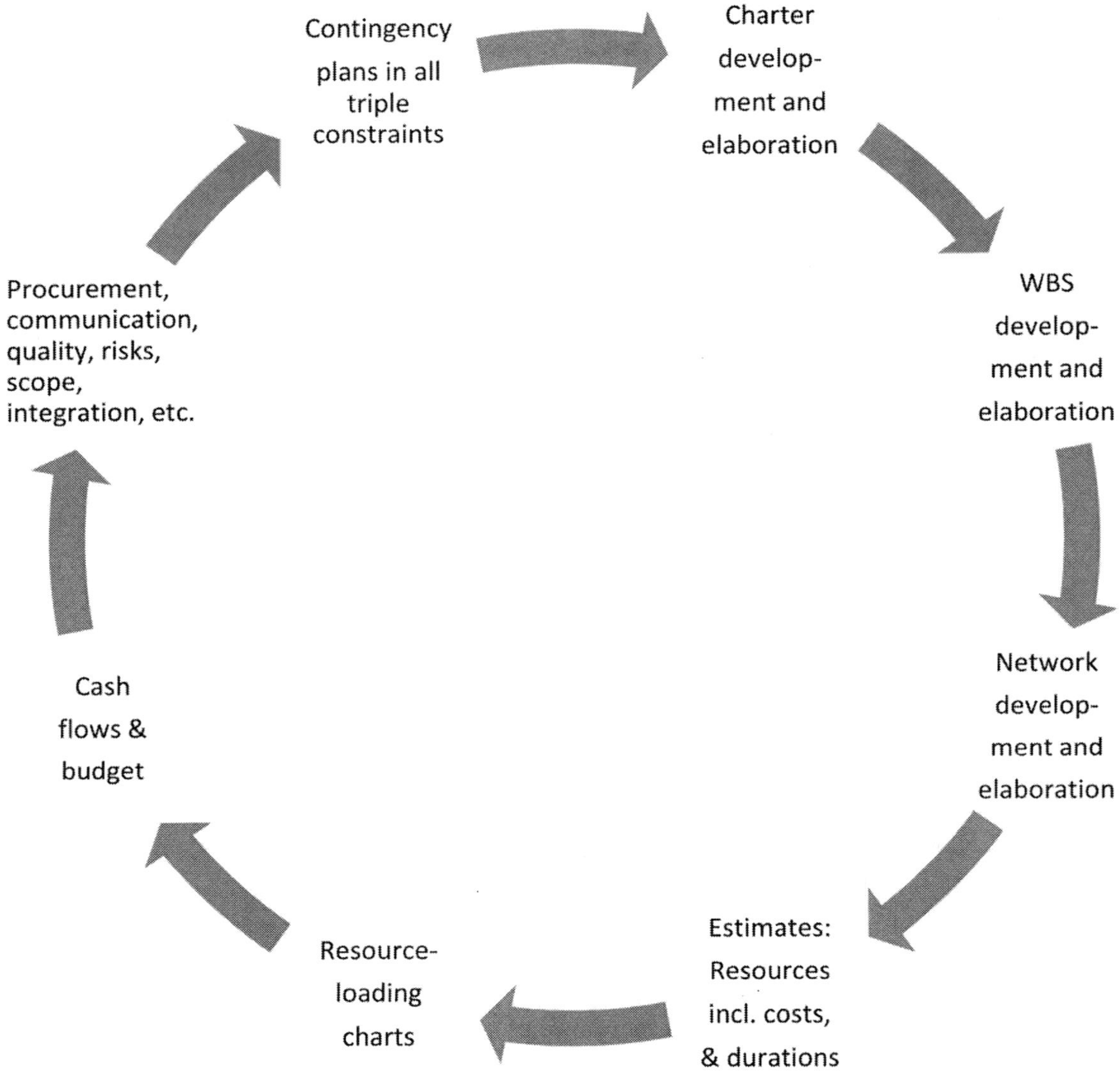

Figure 1. The Project Planning Cycle

495042 Application of Basic Project Management Skills to Small Scale Gas to Liquid Projects

Abstract:

Management of small gas to fuel projects in impoverished communities is an area of interest in Africa. Several projects do not come to completion due to not so proper project planning and execution. With the goal of producing Project managers who are able to complete such complex projects on time and on budget without sacrificing project quality and professionals preparing to transition into a new career in project management we given special hands on training in small gas to liquid projects. Also a group of professionals without formal training who are managing large, complex projects or smaller, more tactical projects were also given the same aid. The training equipped participants with the principles and practices of disciplined project management while addressing the unique challenges of gas to liquid projects. Participants apply the techniques in a small group setting using a realistic case study. Almost 100% gained a comprehensive overview of the skills, knowledge and tools needed to effectively manage projects.

489242 Quantifying Reading and Online Homework Completion Using an Interactive Material and Energy Balances Textbook

Abstract:

Interactive textbooks provide instant feedback to students as well as to the professor. Last year, features of an interactive textbook from zyBooks for a material and energy balance course were detailed, including scaffolded question sets and animations. This year, new features include over 170 automatically graded homework problems and interactive material on using spreadsheets for solving course problems. The machine-graded, personalized homework, which are called challenge activities, include rolling numbers, randomized content, and scaffolded questions across 3 to 6 levels per activity. The increasing difficulty was verified by student success rates in 2017. Quantifying student usage will be central to this contribution. In 2016, student reading/participation averaged 87% over the entire interactive web book over the entire semester with 100 students, which far exceeds the less than 30% reading statistics in the literature. Additionally, statistically significant higher textbook reading scores were observed for students earning A and B final course grades compared to C, D, and F final grades as well as female students compared to male students. Data being generated in early 2017 will be compared to the earlier findings related to course grades and reading.

493310 An Open-Access Gate-to-Gate Life Cycle Assessment for Graduate Researchers

Abstract:

As graduate students become masters of niche areas in narrow sub-disciplines, it is easy for them to lose sight of the bigger picture. Whether itâ€™s the feasibility of the project, or potential impacts, it can be difficult for them to place their work in a larger context. Here we discuss a project-based research assessment for a graduate-level chemical engineering class using resources that may be freely used and adapted. Based off of a simplified Life Cycle Assessment, a materials and energy inventory, this excel based evaluation seeks to quantitatively assess the social, economic and ecological boons and burdens of their own research compared to industrial and academic processes in the literature. Performing cost, toxicity, and environmental analyses allows students to weigh the pros and cons of specific processes, and students learn that few decisions are simple to make. Next students evaluate their work and importantly, suggest possible improvements to the assessment itself. By conducting manageable and practical investigation of their own research, students learn how to contextualize their innovations and understand the relationship between how their choices in the laboratory can result in safer, cheaper, and more benign products and practices. Just as this evaluation is built on open access content and strengthened through iterative student and educator generated advances, simply put, this process is really an opportunity for students to dig into the data, as they mine the literature for a variety of parameters. Using technology to spur collaboration, and whole systems thinking, this project balances what young research would like to do, against what information is available, highlighting next steps. Understanding the nexus of energy, materials and people may seem to be a complex art, but by utilizing available materials to spark inquiry, it can be engineered into a craft.

490487 Chemical Engineering Beyond Politics; A Futuristic World Educational System

Abstract:

Chemical engineering beyond politics; a futuristic world educational system

S. Rohani[1,2], S. Agah[1,3], S. Dini[1,4], E. Hemmat[1], B. Mottahed[1], A. Parvizi-Majidi[1], F. Samadi[1], F. Sefidvash[1]

[1]Baha'i Institute for Higher Education (BIHE)

[2]Department of Chemical and Biochemical Engineering, Western University, Canada, Email: srohani@uwo.ca

[3]Chemical and Biomolecular Engineering, Rice University, TX, USA

[4]Mechanical Engineering Department, Western New England University, Springfield, Massachusetts, USA

We offer an internet-based chemical engineering program beyond borders with no tuition fees to the students and minimum overall costs to the institutions offering the program.

We and several other professionals voluntarily established a chemical engineering department within an online university in Iran called BIHE (Bahá'í Institute for Higher Education).

In early 1980's, following the cultural revolution in Iran, the professors and students in all disciplines who confessed their belief in the Baha'i Faith, were expelled from the universities across Iran. In the following years up to the present time, the Baha'i youth have been denied the right to higher education. International appeals initiated by the worldwide Baha'i community to right this obvious human rights violation did not achieve a meaningful conclusion. Consequently, the Baha'i community in Iran with the help of the then expelled professors and under the guidance of the international administrative body of the Faith, the Universal House of Justice, formed a community-based university, the Baha'i Institute for Higher Education (BIHE) in 1987. As an initiative, *the New York Times* called it "an elaborate act of communal self-preservation, to educate those who were unjustly denied access to higher education"[1]. The origin of BIHE was, thus, rooted in a spirit of purposefulness, and dedication. The belief in the power of BIHE has continued to conduct its courses using accredited professors and in accordance with the highest academic standards.

The unique circumstances surrounding BIHE, whereby its professors teach without compensation and its staff and students participate in university activities at a great personal risk, have forced it since 2005 to combine traditional classroom instruction with online learning to unwittingly become a leader in hybrid education.

Although BIHE issues degrees to all of its graduates, Iranian authorities cannot certify these degrees since the BIHE is not recognized as a degree granting institution. However, even in the absence of such certification, BIHE graduates have been admitted into post-graduate programs in universities across North America, Europe, Asia and Australia. So far, more than 90 universities worldwide have recognized the quality of BIHE courses, and accepted them for credit, allowing BIHE graduates to continue their education abroad[2].

In September 2011, two Nobel Peace Prize laureates, East Timor past President José Ramos-Horta and Archbishop Desmond Tutu, wrote an open letter to the international academic community calling on it to condemn the Iranian government's policies towards the Baha'is, to evaluate BIHE curricula, and consider accepting its credits and to offer online curricula to students in Iran who would otherwise be deprived of the right to higher education[3].

Despite these pressures, the BIHE has been functioning with ever-increasing efficiency. The classrooms and the laboratories of this fledgling educational system are the homes and basements of the Baha'i families. The professors offer their voluntary services with dedication and the students fueled by their love for learning attend regular classes, via internet. There are many graduates of the BIHE who have completed their PhD and higher degree programs in high ranking institutions and currently working as professionals, abroad. A few who upon the completion of their post-graduate degrees chose to return to Iran and serve as faculty members in BIHE, faced imprisonment and harsh treatment by the Iranian government. The BIHE students in various cities in Iran attend classes offered via internet in their home towns and every month travel to Tehran for in-person classes and possible laboratory work. While in Tehran, they are hosted by families who open their homes for this purpose. The classes are offered by the professors via internet with the state-of-the-art technology. The classes are often held in the evenings allowing the students to work during the day.

The BIHE faculty and students in Iran have consistently faced oppressive conditions. However through the efforts of dedicated faculties and staffs in Iran, and an ever-increasing Affiliated Global Faculty from universities around the world, BIHE now offers 38 university-level programs including Chemical Engineering, Chemistry and Industrial Chemistry, Civil Engineering, Architecture, Computer Science, Social Sciences and Humanities, Psychology, and Dentistry.

In 2009, the authors with the invitation and help of a core group residing in Iran launched a new major in chemical engineering. To fulfill the teaching assistant requirement, applied chemistry students who were interested studied some relevant courses intensively in two years along with their own major to serve as teaching assistants. The organizing group reached out to several universities and communities outside and inside Iran to find volunteers to teach chemical engineering courses. Since 2012, a group of 14 individuals from different professional and industrial backgrounds got together to facilitate this process. Since then we have met online each month and consulted about how to best offer the curriculum. The purpose of the chemical engineering program is to offer an outstanding undergraduate degree program and train chemical engineers with excellent technical and leadership skills, integrity, and social responsibility to serve the citizens of Iran and the world.

First year courses consist of General Chemistry 1 and 2 and Chemistry Laboratories, General Mathematics 1 and 2, Physics 1 and 2 and Labs, electricity, Computer Programming, and Humanities.

In Year 2 students take more common basic courses which include Organic Chemistry 1 and 2 and Laboratory, Basic Calculations in Chemical Engineering, Chemical Engineering Thermodynamics 1 and 2, Differential Equations, Fluid Mechanics, and Humanities.

In Year 3 more chemical engineering core courses are taught. These include Mass Transfer, Heat Transfer, Statics, Strength of Materials, Numerical Analysis, Unit Operations 1, Kinetics and Reactor Design, Applied

Math, Introduction to Electrical Engineering, a Technical Elective, and Humanities.

In the final year Unit Operations 2, Chemical Process Control, Modelling and Simulation using software packages such as HYSYS, Introduction to Material Science, Capstone Plant Design, two more Technical Electives and a course in Professionalism and Business are taken.

The technical electives are offered in Waste Water Treatment, Oil and Gas and Petrochemistry, Introduction to Environmental Engineering, Introduction to Biomedical Engineering and a non-technical elective Entrepreneurship for Chemical Engineers at the fourth year level. This provides a unique synergy and integration of courses offered in our program, so that all of our graduates have a holistic appreciation of process and plant design, which includes environmental aspects.

In all courses, emphasis is placed on creativity, innovation, problem solving through teamwork, communications, and reflection – the skills necessary to practice engineering in any discipline.

The design experience in the Chemical Engineering program is realized through the Capstone Design course, as well as the design components existing in the majority of Program Compulsory and Technical Elective courses. The aim is to expose and train students in handling open-ended engineering problems. Solutions must consider safety requirements, codes, standards, specifications, health, environment, social issues, and other pertinent constraints. Students will receive training with regard to economics, business, legal matters, and interpersonal and communication skills. Design courses also provide training in spreadsheets, process simulation, equipment selection, sizing and costing, and cash flow analysis.

We also proposed and managed several research projects under Scientific Advisory Committee since January 2016. This Committee is formed to help facilitate and accelerate the research activities of BIHE that will train our students in conducting research projects and publishing the outcomes in well-known journals.

We praise the students of the BIHE for their ultimate sacrifice and their dedication in pursuit of education. We also offer the operation of the BIHE as a world-wide educational model that can operate at low-cost and high efficiency in many less fortunate countries and remote areas that do not have access to higher education.

[1] http://www.nytimes.com/1998/10/29/world/iran-closes-university-run-covertly-by-the-bahais.html

2 http://bihe.org/index.php?option=com_content&task=view&id=146&Itemid=236

3 http://www.huffingtonpost.com/desmond-tutu/iran-bahai-_b_978090.html

494531 Enhance Learning Experience By Augmented Reality Tools

Abstract:

The increase in the popularity of modern technologies among students is increasingly challenging the status quo of the educational system. Pairing teaching and learning strategies with appropriate technology trends can greatly enhance the motivation to learn. Ergonomics and human factors can play a critical role in analyzing the abilities and behaviors of students and employing the gained knowledge in designing tools, products, and systems that are more effective for accomplishing a given task. Wearable technologies equipped with augmented reality capabilities offer a lot of promise for education and training especially for the applied science disciplines. In this work, we utilize wearable technologies, namely the Google Glass and Microsoft HoloLens, to immerse students in an augmented reality environment that guides actions at every step of the experiment via intelligent audio-visual feedback. In addition, we share our experiences over the tools development and first pilots.

R&D Management from Beginning to End

Jack Hipple
Principal
Innovation-TRIZ
Tampa, FL

Eldon Larsen
Professor of Engineering
Marshall University
South Charleston, WV

Extended Abstract

This workshop focuses upon managing R&D projects from beginning to end. Beginning refers to mental picture of an idea whereas end refers to solving the customers need and in the process generating profit for the organization. Using a case study, the speakers will delineate challenges involved at every stage of R&D project from a mental image of an idea to a profit making product. The talk also compares old paradigm of innovation management with the new proposed paradigm. The probability of success can be significantly improved using the new paradigm.

Successful Innovation Management

We have seen many idea management flow sheets, gate management processes, and prioritization recommendations starting with an idea that already exists, but no discussion of where the idea came from or what was necessary to generate and frame it. Moreover attention is not paid to shaping and propagation of the idea. This has been especially true over the past decade when processes such as Six Sigma came into vogue whose main thrust was to eliminate deviation. New concepts and innovation are, by their very nature, deviant! In order to generate fresh and innovative ideas we must create a new thought pattern that may challenge the existing status quo in any number of ways. These can even include, at an extreme, replacing one's curent products and businesses with totally new approaches and models.

The key elements of successful commercial innovation include:

1. Continuous flow of unmet needs, both from current customers and related to core competencies of the organization ("outside in" as well as "inside out"
2. Using clear mental models of problems and potential solutions
3. Perseverance for idea propagation
4. Clear definition of the business opportunity

5. An appropriate product development and management team
6. Efficient process scale up and continued input from the market
7. Successful project management to insure implementation on schedule and within budget

We will look at each one and discuss the important issues.

1. Continuous Flow of Unmet Needs and Exploration of Uses for Core Competencies

An organization needs to have a process for how it decides what to work on. This may sound simplistic, when this is left to intuition or accident; it is unlikely that the "best" projects will be chosen.

Steps to make sure that an innovation effort is capturing the best opportunities:

- Talk to groups that are not your current customers but that might be able to use the functional capability that you have as a core competence. Customer competitors are another valuable source. The customers of your customers are another excellent source. This insures that you have awareness several levels above your product and customers and alerts you to potential long term needs and trends that may not be obvious to your direct customer or may not be readily disclosed by them.

In these discussions we want to make sure that we have a clear definition of what the "ideal result" that our customers are desiring. We can also look at parallel universes of technology and markets to explore opportunities.

2. Using Clear Mental Models of Problems and Solutions

It is one thing to "invent" a new product, but another to make sure that we understand how a potential customer is going to use the product as well as how the customer' s customer is going to use their product. The days are gone when we can just make something, put it in a drum or tank car, and wait for the invoice to be paid. We must be intimately familiar with the product' s use, its limitations (including regulatory issues if appropriate). We must understand the details of the physical use of the product and how it affects a customer' s products and business.

It is also important to make sure that we do not have blinders on when thinking about use of our product. It is rare that there is only one way to accomplish a function. We must be constantly vigilant about competitive activities, not just from direct competitors, but from parallel approaches to meeting a customer' s needs.

3. The People Side

Ideas, innovation, and projects never happen in a vacuum. They are accomplished with teams of people, all of which have different personalities and styles. Recognizing and using these pro-actively is critical to a project' s success. How these are measure and used will be discussed.

4. The Business Opportunity

A new product is not a business unless we have a way of sustaining the product' s sales, expanding its utility, and developing and enforcing an intellectual property portfolio. As we commercialize a new product, we must be constantly aware of, and be willing to address with resources and commitment, issues such as change in product form, packaging and shipping issues, product purity, and shelf life/stability. These issues are most adequately addressed by constant and continuous communication with the customer.

5. Product Development Teams

We frequently choose product team members based on "who' s available". While this may be expedient, it may not serve the best interests of the new business development. The obvious component of team selection is function—marketing R&D, manufacturing, legal, sales, etc. but we frequently ignore the human side of team functioning. If team members spend too much time dealing with personal issues relating to other team members, that energy is not applied to successful new product launch. Tools that can assist in this include measurements such as the Myers Briggs family of instruments, the Hermann Brain Dominance instrument (HBDI), and the Kirton KAI assessment. Knowledge of these personality profiles will not change natural behavior, but can make team members aware of others' preferred styles and open discussion can eliminate hidden barriers to effective team functioning.

6. Efficient process scale up and continued input from the market

Often times, companies strive for launching the best product. It is important to realize that continued input from the market is critical in order to have that ' best product' in the market. It is necessary to efficiertly scale up and launch a product sooner than later. It should be perfected using market feedback. Getting market feedback is the only way to close the R&D loop and serve your customers to the fullest.

7. Successful Project Implementation and Management

New product development does not end when the first semi-commercial sample has been OK' d by the customer. A new type of team isnow required to oversee the details process design, plant construction (including all the sub issues of permitting, EPA regulations, etc.). Quick and efficient plant start up with appropriate contingency planning is also a part of this effort. Contingency planning, not only related to unforeseen plant construction and start up difficulties, but also to taking care of the customer's needs in these situations must be part of the plan.

Three keys to successful project management lead to successful project completion. These keys involve

- The importance of people
- The importance of defining and setting good objectives
- The importance of good planning

We will apply these three keys to successful project management. This is all about application, not just theory.

498323 A PhD's Perspective on Driving Operational Excellence in a Manufacturing Environment

Abstract:

The manufacture of chlorine and caustic soda, vinyl chloride monomer, and polyvinyl chloride involves handling highly hazardous chemicals. Operational Excellence is essential in helping a facility or a company reduce safety and environmental incidents, improve reliability and product quality, and reduce overall manufacturing costs. A brief review of Chloro-Vinyls technology will be included in the presentation. The general principles, systems and tools for sustainable Operational Excellence in a manufacturing environment, including how to integrate R&D resources into the Operational Excellence process, will be covered. A number of "Best Practices," specific to commodity chemical manufacturing but broadly applicable to any industry, will be highlighted. Past incidents where Operational Excellence would have prevented the incident will be woven into the presentation.

UCOR
URS | CH2M
Oak Ridge LLC

Session 487b

Environmental Cleanup: A Vital Link in the Nation's Nuclear Security Framework

Dr. Harold T. Conner, Jr.
Nuclear Services & Engineering Manager

Safely Delivering DOE's Vision for the East Tennessee Technology Park

UCOR
URS | CH2M
Oak Ridge LLC

About UCOR

The Department of Energy's cleanup contractor for the East Tennessee Technology Park and other sites on the Oak Ridge Reservation

AECOM

- Leader in resolving EM's complex high-risk challenges
- ENR's #1 Top 500 Design Firm
- Fortune 500 company, employing 100,000 people worldwide

ch2m

- Leader in nuclear liabilities management
- ENR's #1 Program Management and Environmental Firm
- Employs 25,000 people worldwide

RSI

- Leader in environmental cleanup and beneficial site reuse
- Small business and minority business enterprise within an Alaska Native Corporation employing 290 people
- DOE National Small Business of the Year (2017)

2

UCOR
URS | CH2M
Oak Ridge LLC

DOE's Oak Ridge Reservation

- Established in the early 1940s as part of the historic World War II Manhattan Project
- 3 distinct sites supporting national defense and energy missions
  - East Tennessee Technology Park
  - Y-12 National Security Complex
  - Oak Ridge National Laboratory

UCOR
URS | CH2M
Oak Ridge LLC

The Work

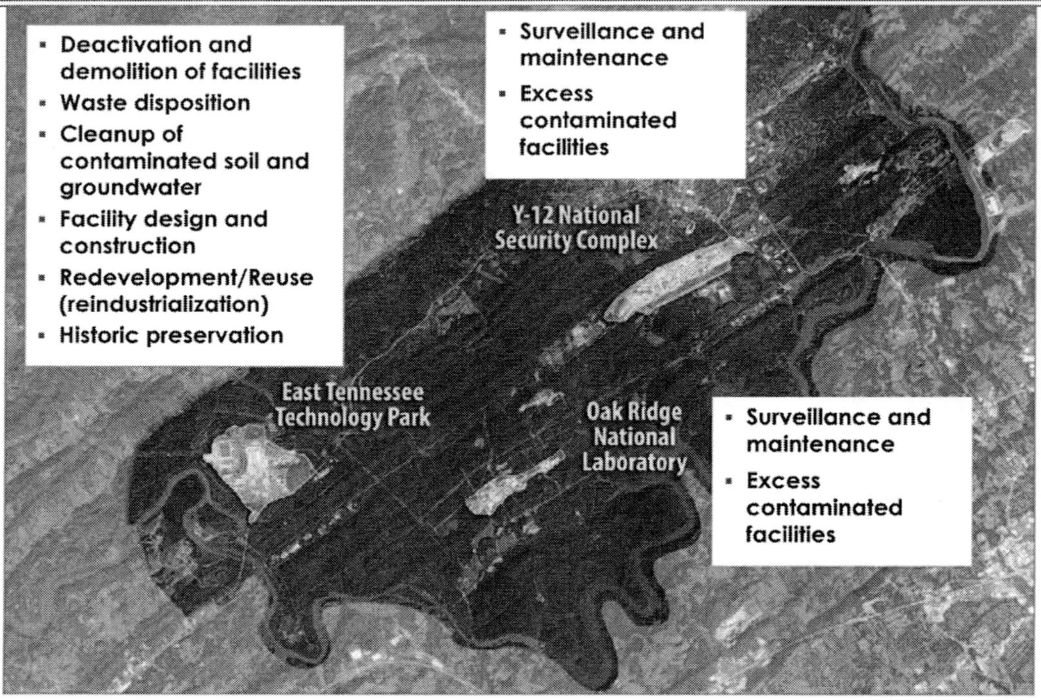

- **Deactivation and demolition of facilities**
- **Waste disposition**
- **Cleanup of contaminated soil and groundwater**
- **Facility design and construction**
- **Redevelopment/Reuse (reindustrialization)**
- **Historic preservation**

- **Surveillance and maintenance**
- **Excess contaminated facilities**

- **Surveillance and maintenance**
- **Excess contaminated facilities**

4

UCOR
URS | CH2M
Oak Ridge LLC

The Hazards

Industrial

Chemical

Nuclear/
Radiological

5

UCOR
URS | CH2M
Oak Ridge LLC

ETTP – Before Cleanup Begins

6

UCOR
URS | CH2M
Oak Ridge LLC

ETTP – August 2016

UCOR
URS | CH2M
Oak Ridge LLC

ETTP - Status

Deactivation

- Cold and Dark
- Characterization including High Risk Equipment removal and mining
- Process gas equipment foaming
- Equipment tabbing/unbolting
- "Go Orange" marking
- Criticality Incredibility determination

Foaming

Compressor and converter unbolting

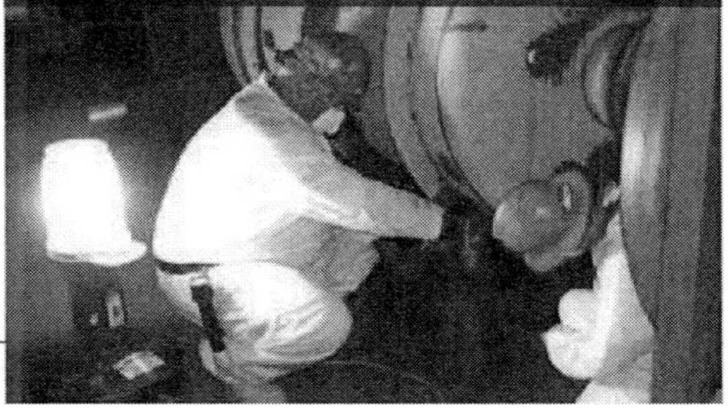

Demolition

- Open-air demolition
  - Fixative and water spray to control dust
  - 14 air monitors to ensure no airborne contamination
- Process equipment/piping with greatest contamination/ material deposits removed from building prior to demolition
- Contents meeting onsite waste acceptance criteria demolished with building
- Storm water pollution prevention plan

First day of K-27 demolition

10

UCOR
URS | CH2M
Oak Ridge LLC

Environmental Remediation

11

UCOR
URS | CH2M
Oak Ridge LLC

Waste Management

12

Vision 2016 Accomplishments

- 4.5 million square feet demolished (5 process buildings)
- 5,000 converters shipped offsite
- 137 miles of piping removed
- 56,000 truckloads of waste shipped onsite
- 1,600 truckloads of waste shipped offsite
- 720 acres transferred

13

Y-12 National Security Complex

14

UCOR
URS | CH2M
Oak Ridge LLC

Y-12 National Security Complex

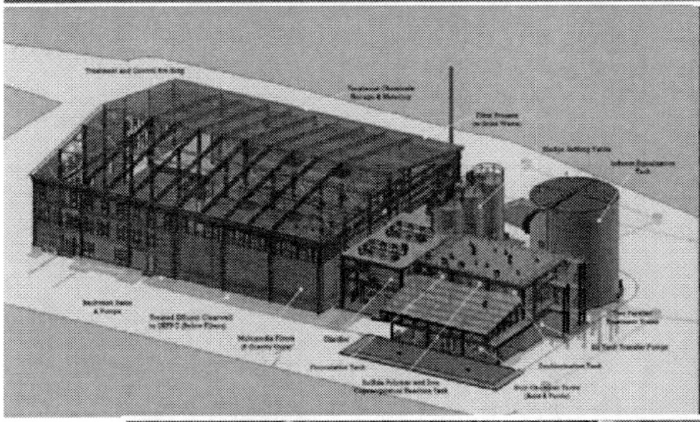

15

UCOR
URS | CH2M
Oak Ridge LLC

Oak Ridge National Laboratory

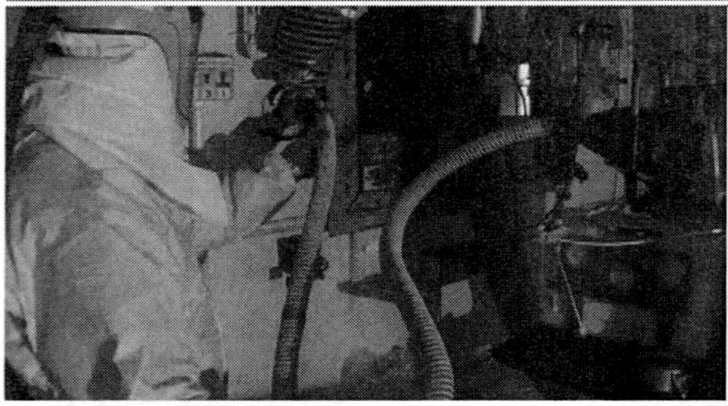

UCOR
URS | CH2M
Oak Ridge LLC

Historic Preservation

17

UCOR
URS | CH2M
Oak Ridge LLC

Redevelopment and Reuse

18

UCOR
URS | CH2M
Oak Ridge LLC

Safe Operations

Safety Excellence

Safety & Health Awards

- National Safety Council **Industry Leader Award** for 9 million safe hours without lost workday injury
- National Safety Council **Superior Safety Performance Award** for 15 years without a lost workday injury at onsite disposal facility, EMWMF
- Verdantix **Environmental Health & Safety (EHS) Innovation Award** for use of drones and other technology to deliver cleanup mission
- Governor's Foundation for Health and Wellness **Tennessee Healthier Workplace** for encouraging and enabling healthier lifestyles
- Voluntary Protection Program Participants' Association **Safety and Health Outreach Award** for providing public opportunities to learn safety skills
- Voluntary Protection Program Participants' Association **Innovation Award** for use of drone to inspect industrial stacks

20

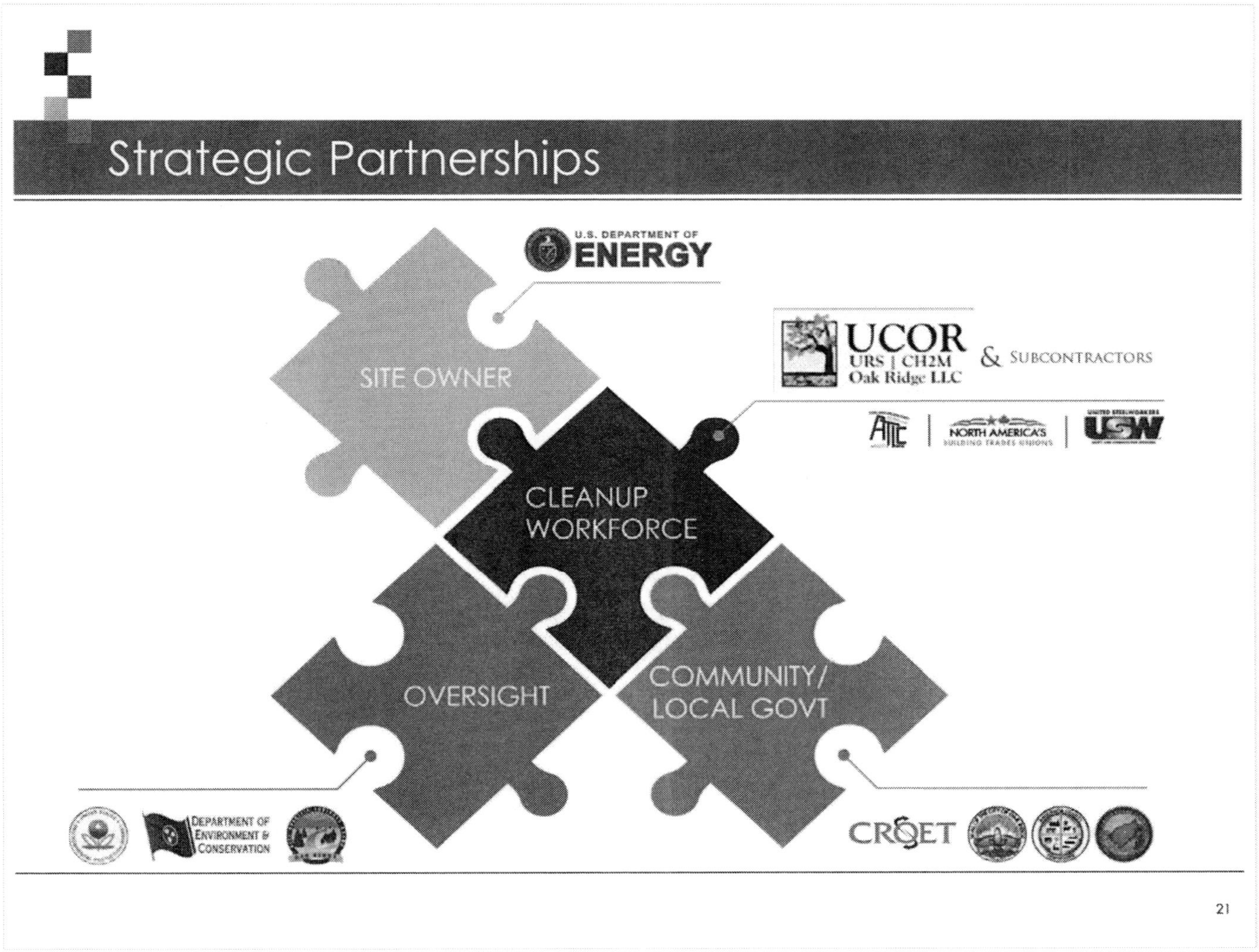
Strategic Partnerships
U.S. DEPARTMENT OF
ENERGY
UCOR
URS | CH2M
Oak Ridge LLC
& SUBCONTRACTORS
NORTH AMERICA'S
USW
SITE OWNER
CLEANUP
WORKFORCE
OVERSIGHT
COMMUNITY/
LOCAL GOVT
DEPARTMENT OF
ENVIRONMENT &
CONSERVATION
CROET
21

487310 Your Career Path Is like a Distillation Column

Abstract:

A typical distillation column will have its distillate stream at the top and residue stream at the bottom. No matter which stage your feed stream is coming from and what the composition og your feed stream is, all the molecules should select a right steam to go.Â Your career path is same. Normally there are two kinds of career paths(managerial career path and technical/professional career path) that you can follow in your career development. No matter what your level of KSA(knowledge, skill and ability) is when you graduate from your college, you should pick up a right career path to follow. Different career path will require different KSA development so it is better for you to make a decision as early as possible. This presentation will provide youâ€™re the detailed comparison of two career paths and your development approach of required KSA.

AUTHOR INDEX